OECD Studies on Water

Water and Cities

ENSURING SUSTAINABLE FUTURES

OECD
BETTER POLICIES FOR BETTER LIVES

This work is published under the responsibility of the Secretary-General of the OECD. The opinions expressed and arguments employed herein do not necessarily reflect the official views of OECD member countries.

This document and any map included herein are without prejudice to the status of or sovereignty over any territory, to the delimitation of international frontiers and boundaries and to the name of any territory, city or area.

Please cite this publication as:
OECD (2015), *Water and Cities: Ensuring Sustainable Futures*, OECD Studies on Water, OECD Publishing, Paris.
http://dx.doi.org/10.1787/9789264230149-en

ISBN 978-92-64-23010-1 (print)
ISBN 978-92-64-23014-9 (PDF)

Series: OECD Studies on Water
ISSN 2224-5073 (print)
ISSN 2224-5081 (online)

Co-published by IWA Publishing
Alliance House, 12 Caxton Street, London SW1H 0QS, UK
Tel: +44 (0)20 7654 5500, Fax: +44 (0)20 7654 5555
publications@iwap.co.uk
www.iwapublishing.com

ISBN: 9781780407593 (Paperback)
ISBN: 9781780407609 (eBook)

The statistical data for Israel are supplied by and under the responsibility of the relevant Israeli authorities. The use of such data by the OECD is without prejudice to the status of the Golan Heights, East Jerusalem and Israeli settlements in the West Bank under the terms of international law.

Preface

Cities are major contributors to national economies and play a key role as nodes in global markets. But cities can only develop sustainably when they provide reliable water supply and sanitation services to city dwellers and manage risks of too much, too little or too polluted water.

In OECD countries, cities have achieved high levels of protection against droughts, floods, or water pollution, and a vast majority of city dwellers enjoy reliable water services. This remarkable performance derives from distinctive combinations of infrastructures, business models and institutional arrangements. However, whether and how such combinations are fit for future challenges is unclear. The economic, social and environmental costs of water security are increasing, driven by urban growth, competition among water users, urban and agricultural pollution, and climate change. Existing infrastructures are also ageing and need to adapt to new contexts. In addition, city dwellers have rising expectations as regards the quality of water services and water security.

Can these challenges be addressed with the current business models and financial resources? What are the opportunities that derive from innovative technologies and practices? How can we tap into the potential benefits of co-operation between cities and their rural environment? What governance arrangements are required? This report provides answers to all these questions. *Water and Cities: Ensuring Sustainable Futures* argues that cities in OECD countries face significant revisions of financial, technological and governance arrangements. While some cities have already gained experience with managing this transition, more needs to be done to scale up and expedite change. Both local and central governments have a role to play in order to use to the best advantage the initiatives of a variety of stakeholders, including the private sector, households and rural communities.

Urban water management is a domain where a dialogue between developed and developing countries is most promising. Fast-growing cities in developing countries can experiment innovative urban planning and water management practices, leapfrogging arrangements that have locked-in OECD cities in inflexible practices. At the same time, OECD cities can learn from these developments and explore how they can be inserted in existing infrastructures and institutional arrangements.

This report builds on a wide range of expertise across the OECD. It is a good illustration of how cross-sector co-operation can help identify emerging issues, generate new knowledge and pave the way to innovative policy responses. I am confident that *Water and Cities: Ensuring Sustainable Futures* will inspire more innovative urban water management practices, at different scales and in a variety of contexts.

Angel Gurría

Secretary-General, OECD

Foreword

OECD cities usually benefit from high levels of protection against water risks and most city dwellers have access to reliable water services. At the same time, OECD cities face significant challenges to protect inhabitants from risks of floods, droughts or deteriorating water quality resulting mainly from urban growth, competition among water users, urban and agricultural pollution and climate change. They also face particular challenges due to ageing infrastructures and the need to adapt existing assets: most OECD cities need to transition from an era of exploiting existing infrastructures to one of building new assets and inserting such assets in existing environments.

This report focuses on the challenges confronting OECD cities in terms of water management and explores policy responses at both the central and local government levels. The analyses focus on four mutually dependent dimensions: finance, innovation, co-operation with the rural environment and governance.

The report builds on OECD's work on water, particularly on financing water management and water services; the diffusion of technical and non-technical innovation in water management; the management of droughts, floods and groundwater; allocation of freshwater; urban and multi-level governance; stakeholder engagement; and the governance of water regulators.

The report was drafted by a core team comprising Xavier Leflaive, Aziza Akhmouch, Filippo Civitelli, Tatiana Efimova, Guillaume Gruère, Julien Hardelin, Celine Kauffmann, KunWook Kim, Hannah Leckie, Kazuki Motohashi and Oriana Romano. Simon Buckle, Anthony Cox, Jane Ellis, Robert Youngman, William Tompson, and Karishma Gupte provided comments at various stages. Romy de Courtay edited an earlier draft and Peter Vogelpoel formatted it. Sama Al Taher Cucci seamlessly managed the administrative process.

The report builds on case studies prepared by selected OECD cities: Auckland (New Zealand), Fukuoka (Japan), Hamburg (Germany), San Francisco (United States), Suwon (Korea), Tokyo (Japan), Tucson (United States) and four local authorities in South West Gyeongnam Province (Korea).

It also builds on new information collected through a survey on the governance of water regulators and a survey on water governance in cities. The outcomes of each survey will be published in separate companion reports.

Table of contents

Figures

Tables

Boxes

Abbreviations

ATSE	Academy of Technological Sciences and Engineering
AWS	Assured water supply
BMA	Barcelona Metropolitan Area
CAP	Central Arizona Project
CBFR	Consumption-based fixed-rate water rates
CER	Commission for Energy Regulation
CGE	Computable general equilibrium
CPUC	California Public Utilities Commission
CWRS	Comprehensive Water Reuse System
DEP	Department of Environmental Protection
EASAC	European Academies of Sciences Advisory Council
EPA	Environmental Protection Agency
EU	European Union
FOEN	Federal Office for the Environment
GDP	Gross domestic product
GIS	Geographical information systems
GWI	Global Water Intelligence
HW	Hamburg Wasser
IBT	Increasing block tariffs
ICT	Information and communication technologies
IFM	Integrated flood management
ITU	International Telecommunication Union
IWNL	Independent Water Networks Limited
MCBA	Modified cost balancing accounts
MGSDP	Metropolitan Glasgow Strategic Drainage Partnership
MHLW	Ministry of Health, Labour and Welfare
MNB	Marginal net benefits
MOU	Memorandum of understanding

NRC	National Research Council
O&M	Operations and maintenance
OMR	Operation, maintenance and rehabilitation
PES	Payment for ecosystem services
PSP	Private-sector participation
RISA	Rain InfraStructure Adaptation
S&R	Storage and recovery
SFPUC	San Francisco Public Utilities Commission
SuDS	Sustainable drainage systems
SWI	Showcasing Water Innovation
SWM	Smart water management
TAD	Trade and Agriculture Directorate
USACE	US Army Corps of Engineers
USEPA	US Environmental Protection Agency
WBRS	Water budget rate structure
WMO	World Meteorological Organization
WPBWE	Working Party on Biodiversity Water and Ecosystem
WRAM	Water revenue adjustment mechanism
WSIP	Water Savings Incentive Program
WSS	Water supply and sanitation

Executive summary

Cities in OECD countries have not solved water management. While they currently enjoy relatively high levels of protection against water risks, they face disquieting challenges, including the proven difficulty of upgrading and renewing existing infrastructures, and heightened uncertainty about future water availability and quality. Cities in OECD countries are entering a new era, characterised by the need to retrofit existing assets into more adaptable infrastructure, by different combinations of financing tools and by new roles for stakeholders in water management. The transition to this new era requires co-ordinated action among central governments, local authorities and a variety of private actors.

Emerging challenges for water management in OECD countries

City dwellers in OECD countries currently have reliable access to safe water and sanitation services. They are protected against risks pertaining to floods, droughts, water pollution and reduced resilience of freshwater ecosystems. This remarkable achievement builds on specific combinations of institutions, financing mechanisms and technologies. Yet such combinations may not be fit for emerging challenges.

One of the main challenges is the increased uncertainty about future water availability. First, cities compete with other water users (farmers, energy suppliers and the environment) to access the water they need at fit-for-purpose quality. Second, climate change will generate more extreme weather events, increased hydrological variability and higher uncertainty about water availability.

Another challenge is the financing of the operation, maintenance and renewal of extensive infrastructures that channel, store, treat or move water. Prevailing financing mechanisms have generally been able to support the operation of existing infrastructure, while keeping water tariffs relatively low. They have been less successful at financing the upgrade or replacement of assets, nor have they provided incentives to adapt urban water management to changes in water availability or consumption patterns.

A third challenge is water governance. Territorial reforms and the reallocation of competences affect urban water management, in the same vein as emerging changes in regulatory models for the water industry. Urban water management in OECD countries suffers from several governance gaps, notably fragmented institutions, weak capacity at the local level and tensions between water, energy and land policies.

Four questions to set urban water management on a sustainable path

In this context, cities in OECD countries would benefit from considering four interrelated questions.

What should the water bill cover? The OECD has long argued that three ultimate sources of financing (the 3Ts) exist for water services: tariffs (revenues from the water bill), taxes (allocations from the public budget) and transfers from the international community (which have become secondary in most OECD countries). Other sources of funding need to be paid back from a combination of the 3Ts. The context outlined above suggests a revision of the 3Ts is necessary. Tariff structures and business models in particular may need adjusting in order to secure stable revenues in the face of declining water consumption. Governments should consider levying taxes on those (including land and property developers) who benefit from increased water security or who generate higher costs and externalities (e.g. owners of large impervious surfaces, such as roads or car parks).

How can cities make the best use of innovative approaches to urban water management? Technical innovation is flourishing, but is not fully exploited. Some innovation, such as smart technologies, distributed systems or green technologies, is potentially disruptive. Disruptive technologies work best in combination with non-technical innovation, such as water-sensitive urban design or innovative business models for water utilities. Cities would benefit from having a wide latitude to explore technologies that fit local contexts. Regulatory frameworks can drive the diffusion of innovation, but can also lock cities into sub-optimal technical trajectories.

How can cities and their rural surroundings best co-operate? The urban-rural interface can contribute a great deal to protecting cities against water risks now and in the future, at least cost to society. For instance, experience with catchment protection from harmful agricultural practices, or the use of farm land as buffer against floods, has highlighted the efficacy of innovative measures for urban water management. On the other hand, rural communities can use a city's run-off, treated wastewater and nutrients recovered. National governments should provide incentives and institutional mechanisms to foster the use of co-operative arrangements benefiting cities, upstream and downstream communities, and ecosystems.

How can cities govern urban water management? Three issues deserve particular attention. The first issue is stakeholder engagement: such questions as the appropriate level of water security for a city, how much city dwellers are willing to pay for it and how far they are willing to adjust their behaviour can only be answered in practice, when stakeholders are properly involved in decision-making and implementation. Where they have been established, water regulators can significantly contribute to improved urban water management through greater transparency and improved credibility of decision-making and accountability to users. Finally, water management will increasingly take place over a range of scales – from basin to catchment to individual buildings – depending on the particular service required (protection against floods or droughts, water supply, sanitation, drainage, etc.), technological or behavioural sophistication. How can these different scales and levels of decision-making be integrated for maximum mutual benefit? A mechanism consistently used in cities in OECD countries is metropolitan governance, which offers the ability to combine the different scales and pool financial and technical resources across municipalities in a metropolitan area.

Lessons learnt from the transition of selected cities towards a new era

Most cities in OECD countries have core competences to manage water to meet future challenges: they are generally responsible for land use, construction and buildings, and natural resource management. They are well positioned to develop solutions to hydrological, climatic, social or economic conditions. They can catalyse action by households, local communities and investors.

Some cities have transitioned towards forward-looking water management practices. Their practical experience with managing change and retrofitting existing infrastructures is particularly instructive. Cities in OECD countries that manage water for future challenges understand that delaying action can increase costs and limit options to adapt to new water-related risks. They combine a long-term strategy with a pragmatic approach to renewing the stock of buildings and assets. They deploy a package of technical and non-technical measures that exploit existing water resources, financial capacities and various stakeholder initiatives.

Cities in OECD countries will not be in a position to respond to all the future water challenges on their own. Initiatives by other tiers of governments, clustered around three categories – regulation (on land use, reclaimed water or public procurement), resource provision (e.g. information and education) and incentives (e.g. awards) – will also contribute to urban water management. Governments can use urban policies and infrastructure financing to promote water-sensitive urban design, especially in high-risk regions.

The interplay between national and local initiatives on water management will shape the cities of the future, including their capacity to thrive and contribute to bettering the lives of their residents.

Chapter 1

A framework for city-level water management

This chapter examines the main water-related challenges facing cities in OECD countries now and in the future. Central to these challenges are the risks associated with water abundance, water scarcity, water pollution, water ecosystem resilience and the distinctive ways in which cities in OECD countries have managed these risks so far through infrastructures and governance. The chapter proposes a framework to analyse policy responses to these challenges, combining four dimensions: financing, innovation, urban-rural interface and governance. Subsequent chapters explore each dimension further.

Key messages

Cities in OECD countries enjoy high levels of water security, and city dwellers have access to reliable water supply and sanitation (WSS) services, contributing to public health, economic activity, environmental ecosystem protection and human well-being. This remarkable achievement builds on country-specific combinations of institutions, financing mechanisms and technologies.

However, the prevailing combinations may not be up to future challenges:

- Infrastructures, which underlie water security cities in OECD countries, are ageing and require upgrading. The prevailing business models for urban water management fail to attract needed financial sources.
- Urban water management faces emerging pressures, such as more-stringent health and environmental standards, diffuse pollution, competition to access the resource, increased intensity and frequency of extreme weather (affecting precipitation and evaporation) and more generally, higher uncertainty about future water availability and demand. Cities in OECD countries, however, are locked-in technical trajectories, and retrofitting existing infrastructure is particularly challenging.
- Water governance in cities is hindered by several gaps, such as information asymmetries, sectoral fragmentation and limited capacities. Moreover, institutional structures are changing, driven by national and international laws and regulations, territorial reforms, decentralisation and the reallocation of competences across jurisdictions. These changes affect the capacity of cities and other actors to manage water at the appropriate scale.

Responses to these challenges combine four dimensions: financing mechanisms and strategies to upgrade and renew existing infrastructures; policies to overcome barriers to the diffusion of innovative approaches to urban water management; enhanced co-operation between cities and their rural surroundings; and governance structures that can manage water at several scales, engage stakeholders and properly regulate WSS services.

Some of these responses will be generic, e.g. developing a long-term vision to guide short-term and piecemeal developments; stimulating initiatives by local entrepreneurs and various stakeholders; and addressing equity issues arising from water management. Others are more appropriate to specific issues for each individual city: demand management is more appropriate in a water-scarce environment, while distributed water systems are more competitive when existing networks have reached their full capacity or reclaimed water can be used locally.

This report proposes a typology based on urban characteristics that shape the appropriate response options for a given city in the face of different challenges. This typology presents three main dimensions: the prevailing water resource (surface versus groundwater); urban features (e.g. affluence, spatial patterns, or urban dynamics); and the institutional architecture (i.e. fiscal autonomy and the co-ordination mechanisms with surrounding cities and territories).

Cities will not respond to future water challenges on their own. A number of initiatives taken by other tiers of governments will contribute to urban water management, clustered here around three categories: regulation (on land use, reclaimed water and public procurement), resource provision (e.g. information and education) and incentives (e.g. financial). Future water management in cities in OECD countries will largely depend on the capacities of different tiers of government to work together along a coherent pathway, as well as engage with and make the best use of initiatives by local entrepreneurs and stakeholders.

Introduction

Cities in OECD countries are critical to national economic, social and environmental performance. About two-thirds of the population of OECD countries live in cities – which account for an even larger share of economic output. Cities are hubs for job creation, innovation, growth and culture. Yet they are also sources of acute policy challenges for governments, including infrastructure bottlenecks, high pollution levels, climate and environmental change, and difficulties in the provision of key services (OECD, forthcoming a).

Water security[1] is essential for cities to continue to thrive and contribute to national (and global) economies. Cities in OECD countries are usually well-equipped to handle water-related risks: they are protected from floods and droughts, they can treat polluted water, and a vast majority of city dwellers enjoy premium water and sanitation services. When exceptional events do occur, cities are generally able to overcome and recover from them.

Cities in OECD countries have achieved the prevailing level of water security through a combination of technical and institutional features (OECD, 2013a; Sitzenfrei and Rauch, 2014) which emerged over the 19th century (see Box 1.1):

- Water infrastructure and water polices focus on water quality and supply, with little or no emphasis on water demand.
- Water security relies on technologies to augment supply, as well as treat polluted water sources and effluents.
- Water is collected, distributed and treated in large infrastructures comprising pipes, pumps, taps and meters, centrally organised at the city level.
- Water operators charge users for the service at below full-cost recovery levels.
- Cities are (relatively) disconnected from their rural environment with regard to water management.

This model was sustainable as long as water demand and revenue continued to increase, and new water supplies could meet the growing demand. Public authorities and operators relied on technical innovation to control water pollution. They knew little about making or managing demand and had no relationship with their customers beyond billing. Today, three sets of challenges – water-related risks, ageing infrastructures and trends in institutional reforms – are calling this model into question. The chapter reviews these challenges sequentially and outlines a framework for organising policy responses. It proposes criteria cities can use to compare their water management policies and practices with those of other cities.

Water-related risks to cities in OECD countries

As defined by Grey and Sadoff (2007), water security is "the availability of an acceptable quantity and quality of water for health, livelihoods, ecosystems and production, coupled with an acceptable level of water-related risks to people, environments and economies". Similarly, the OECD (2013b) defines water security as the management of four water risks: scarcity, floods, pollution and freshwater ecosystem resilience. This section reports on recent trends affecting these risks.

Box 1.1. **A brief history of prevailing urban water management models**

The 19th century saw a radical transformation in urban water management. The progressive generalisation of water taps and meters allowed for quantifying water use, and made it possible to distinguish between the use of water as a consumer good or a resource. The invention of bacteriology, and subsequently of chemical water treatment, supported a further shift in the status of safe water from a club good to a payable public good.

While European and other developed countries have largely implemented this transformation, a wide variety of financing systems co-exist. Economic theory argues that water and water-related services should be priced, and that prices should reflect long-term marginal costs. In practice, however, places remain – including cities – where water supply costs are covered by flat rates, sewer services are charged through taxation, or metering is collective (one per condominium) or non-existent.

At the end of the 20th century, the prevailing model for urban water management in developed countries was based on supply-side policies and cost recovery, with minimal subsidies provided (once the main infrastructures had been built). Urban water was treated to potable standards, humans ingested less than 1% of the water supplied, and less than 20% of the water was in contact with human bodies. (Paris, which has a dedicated network for non-potable water – e.g. for cleaning streets – is an exception.) This could be viewed either as excessive water treatment, or on the contrary as the sensible use of economies of scale – treating all the water in a city to the same standard – and application of the precautionary principle – avoiding mistakes that could lead to consuming unsafe water.

The invention of both water and sewage treatment works gave additional degrees of freedom to cities towards their environment: water abstracted from polluted resources could be made potable, and the treated water discharged could have reduced impacts on downstream resources and users. It was only a matter of cost, and for a long time funding WSS services through water bills, with the support of some mutual funding mechanisms and government subsidies, seemed possible.

Source: Barraqué and Isnard (2013).

Water scarcity

Cities are projected to compete increasingly with other water users to access the water they need. In OECD countries, future water demand will be driven by energy suppliers and industry, competing with cities, farmers and ecosystems for access to water (Figure 1.1). In contested basins, competition will intensify as cities continue to expand (United Nations Department of Economic and Social Affairs [UNDESA], 2009, 2010). Competition will extend further as cities fetch water from ever-distant surroundings. Jenerette and Larsen (2006) determined that the land area required to supply water to cities with over 750 000 inhabitants has grown by 28% between 1950 and 2015, at increasingly rapid rates. European cities stand at the lower end of the scale, but as European territory is particularly urbanised, the urban water supply actually affects the entire European territory (Jenerette and Larsen, 2006; Bolognesi, 2014). Competition will intensify further as stakeholders realise that basins have their own needs, and cities must pay a higher economic, financial or political cost to meet their needs. For instance, in the western United States, long-haul water transfers from one source to a distant city met with resistance from local communities (Postel, 1997), with the result that future water-supply plans reverted to local water options.

Climate change has the potential to exacerbate competition over access to water, especially in mid-latitude and sub-tropical dry zones, or where mountain glaciers store less water than they used to. The OECD survey on water governance in cities (OECD, forthcoming b) acknowledges climate change as the major socio-economic and environmental factor that will affect water in cities in OECD countries. This applies to both cities that rely on surface water and cities that rely on groundwater, if they face salinisation (e.g. in coastal areas). In fact, groundwater depletion may become the greatest threat to urban water supplies in several regions in the coming decades (OECD, 2012a).

Cities in several OECD regions already experience severe or recurrent scarcity.[2] For example, the number of areas and people affected by droughts in Europe rose by almost 20 %, at a cost of EUR 100 billion (euros), between 1976 and 2006 (European Commission, 2012). In Mexico, increasingly severe droughts are projected to occur in cities located in the Central, Jalisco and Chiapas regions. In Australia, drought frequency will fluctuate from -20% to +80% by 2070 relative to present conditions, depending on the location; the largest increase in drought frequency is projected to take place in southwest Western Australia (OECD, 2013c). In the United States, half of the 50 fastest-growing cities are located in the drought-prone South. Atlanta, the fastest-growing metropolitan area in the United States, nearly ran out of water in 2007 (Gerrity and Snyder, 2011) and is competing with the seafood industry to secure access to water. In California, global warming and warmer winters are reducing the amount of snow stored in the mountains and causing snowpack to melt earlier in the spring. The result is a persistent and exceptional drought (Brewer, 2014) and water emergencies caused by water shortages in reservoirs, streams and wells.

Figure 1.1. **Projected future water demand**

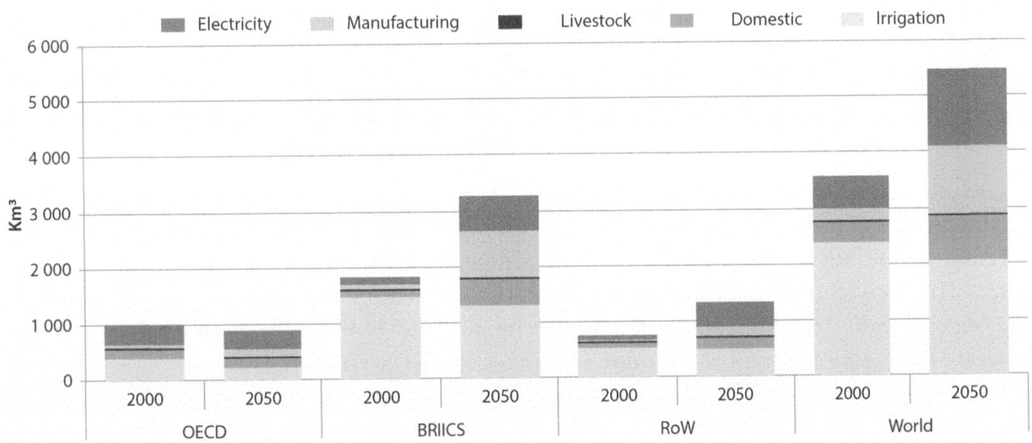

Source: OECD (2012a), *OECD Environmental Outlook to 2050: The Consequences of Inaction*, http://dx.doi.org/10.1787/9789264122246-en; output from IMAGE.

Analyses of water stress in cities in OECD countries need to factor in infrastructure and traditional adaptation to water shortage. McDonald et al. (2014) argue that hydrological models tend to overestimate cities' water stress because they fail to account for canals (to move water from abundant to water-stressed regions), reservoirs (to store abundant water for future use) and groundwater use. Four detailed case studies developed by Richter et al. (2013) illustrate the importance of infrastructure to the urban water supply (Box 1.2).

Box 1.2. Cities' reliance on infrastructure in OECD countries

Richter et al. (2013) document the sequence of policies adopted by cities in Australia (Adelaide) and the United States (Phoenix, San Antonio and San Diego) to secure access to water. The case studies highlight striking similarities and successive strategies to secure the water supply:

- Cities initially tapped their local water supply until it became exhausted. The authors found numerous cases in which cities shifted from a groundwater to surface water supply (or vice versa) when the initial resource became heavily depleted, building reservoirs to exploit extensively the local water resources.
- Next, cities turned to inter-basin water transfers. However, towards the end of the 20th century, the financial, environmental and social consequences made water conservation (a low-cost, low risk option) more appealing.
- In the 1980s, cities started investing in water conservation, a trend that has since gained traction.
- While today's option portfolios include desalination, its still-marginal development remains hampered by financial, energy and environmental constraints.

Source: Richter, B.D. et al. (2013), "Tapped out: how can cities secure their water future?", *Water Policy*, Vol. 15, pp. 335-36.

McDonald et al. (2014) were able to measure urban reliance on infrastructure to secure water. Building on innovative data combinations, they calculated that the world's largest cities draw water from almost half of the global land surface and transport it over a cumulative distance of 27 000 kilometres (km). Los Angeles is a case in point, pulling large amounts of water from sources located outside of its watershed (Postel, 1997). When infrastructures to access water are factored in, 25% of the world's largest cities (above 750 000 inhabitants) are water-stressed (McDonald et al., 2014).

It follows that cities face two combined limitations to accessing water: geographical (the location, quantity and quality of the nearest water sources) and financial (the capacity to reach potentially distant and/or polluted water sources). The first limitation calls for water efficiency and co-operative arrangements to allocate water where it is most needed. The second calls for innovative financial schemes and potentially economic growth to generate the resources required for cities to invest in infrastructures.

Floods

Floods are one of the most costly and damaging disasters, which will pose a critical problem to city planners as they increase in frequency and severity. The latest floods (e.g. Genoa 2014, northern England 2012, New York City 2012, and New Orleans 2005) – caused by sea-level rise, increased precipitation, inland floods, and frequent and violent storms and cyclones – have been very costly to exposed cities dealing with physical damages, human losses and losses from economic activities. Climate change and urbanisation are increasing the risk of urban flooding caused by both rainfall and sea-level rise.

The *OECD Environmental Outlook to 2050* (OECD, 2012a) projects that the number of people at risk from floods will rise from 1.2 billion today to around 1.6 billion in 2050 (nearly 18% of the world's population). The economic value of the assets at risk is expected to amount to approximately USD 45 trillion (US dollars) by 2050, an increase of over 340% from 2010. This conservative estimate does not factor in the effects of climate change. By region, the economic value at risk will increase by 130% for OECD countries, over 640% for Brazil, Russian Federation, India, Indonesia, China, South Africa (BRIICS) and nearly 440% for developing countries.

Cities concentrate a large share of these risks and consequences. Some are located in coastal zones or flood-prone areas; others will be adversely affected by heavy rains, straining the drainage systems. Sealed roads and pavements in cities also alter run-off from rain and stormwater, impair the recharge of groundwater aquifers and increase flood risks.

Building on a new set of data, Hallegatte et al. (2013) assessed and projected financial losses from floods in the 136 largest coastal cities. In 2005, average global flood losses in these cities were estimated at around USD 6 billion per year. Some cities in OECD countries – Miami, New York, New Orleans, Nagoya, Tampa, Boston, Osaka-Kobe and Vancouver – top the list. Due to their high wealth and relatively low flood protection level, three US cities (Miami, New York City and New Orleans) today account for 31% of the losses across the 136 cities.

The costs can be high: shoreline retreat in the United States is projected to cost between USD 270 billion and USD 475 billion for each metre of sea-level rise (OECD, 2010). However, the cost of flood protection is minimal compared with the cost of inaction: for example, flood management in the Netherlands – where more than half of the territory and population and two-thirds of the economic activity are flood-prone – totals USD 2.0 to USD 2.7 billion a year, i.e. 0.3% of the gross domestic product (GDP) (Delta Programme, 2013).

Pollution

OECD countries have made significant progress in reducing pollution loads from municipal and industrial point sources by installing wastewater treatment plants and reducing chemical use. As a result – and despite population increases – nutrient levels in wastewater discharge are projected to remain relatively stable in Canada, Mexico, the United States and European OECD countries (Figure 1.2).

However, while improvements in freshwater quality are easy to discern from point sources (i.e. wastewater discharges), pollution loads from diffuse agricultural and urban sources (i.e. fertilisers and pesticides, run-off from sealed surfaces and roads, and pathogens and pharmaceuticals in animal and human waste) pose persistent challenges in many countries (OECD, 2012a). In nearly half of OECD countries, nutrient and pesticide concentrations in both surface and groundwater in agricultural areas exceed national recommended limits for drinking water standards. This poses a risk to human and environmental health and increases costs for urban water treatment.

Water pollution is affected by shifts in water quantity. Floods decrease the performance of stormwater treatment (including sewerage overflow) and pollutants from different surfaces (e.g. heavy metals from road run-off) contaminate stormwater discharge (Mikovits et al., 2014). Droughts decrease the dilution potential of contaminants in point-source discharges to surface waters, and additional treatment of wastewater may be required to compensate for the lower dilution capacity of the water bodies. In some coastal areas, sea-level rise associated with climate change also increases the risk of saltwater intrusion into groundwater, which can limit its use for municipal or irrigation purposes.

Figure 1.2. **Nutrient effluents from wastewater**

1970-2050

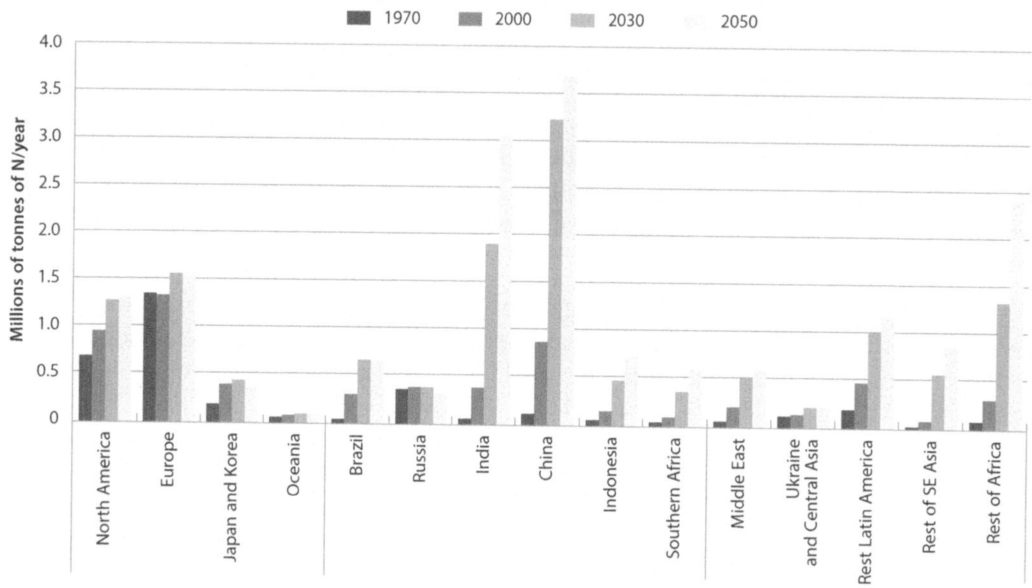

Source: OECD (2012a), *OECD Environmental Outlook to 2050: The Consequences of Inaction*, http://dx.doi.org/10.1787/9789264122246-en; output from IMAGE.

The economic costs of treating water to remove nutrients, pathogens, pesticides and persistent micro-pollutants in order to meet drinking water standards are significant in some OECD countries. So are the consequences of inaction, e.g. outbreaks of waterborne diseases (Box 1.3). The expected increase in the frequency and intensity of extreme weather events, and the high river-flow rates caused by climate change, may cause re-suspension of pollutants stored in sediments, exacerbating the problem. The challenge of treating water to remove contaminants is compounded by growing public expectations of infrastructure performance and human and environmental protection.

A growing source of concern is emerging pollutants and their effects (known or unknown) on aquatic ecosystems and human health (OECD, 2012a). The concept of emerging

Box 1.3. **The costs of drinking unprotected water**

In 1993, the largest recorded waterborne disease outbreak in the United States occurred when treatment plants in Milwaukee (Wisconsin) failed to eliminate Cryptosporidium oocysts introduced in surface waters by run-off from nearby cattle pastures. The incident resulted in more than 4 000 cases of illness (25 % of the population) and 104 deaths in just 2 weeks. According to an analysis by the Centers for Disease Control, the total cost associated with the outbreak was USD 96.2 million (1993 dollars), including USD 31.7 million in medical costs and USD 64.6 million in productivity losses (Corso, 2003). Note that these estimates provide only a lower bound on the true economic cost of the outbreak, since they do not consider willingness to pay to avoid the deaths and illnesses caused by the outbreak.

Source: US Environmental Protection Agency (USEPA) (2013), *The Importance of Water to the U.S. Economy. Synthesis report*, USEPA, Washington, DC, http://water.epa.gov/action/importanceofwater/upload/Importance-of-Water-Synthesis-Report.pdf.

pollutants is not clearly defined; it broadly covers new materials (e.g. nanomaterials), new chemicals and existing chemicals for which new adverse effects are still being identified. Emerging pollutants have a variety of uses in human and veterinary pharmaceuticals, cosmetics, cleaning agents, pesticides and biocides. Persistent or acutely toxic micro-pollutants in water bodies can generate adverse health and environmental effects, such as interference with the endocrine (hormone) systems of humans and animals leading to cancers, birth defects and other developmental disorders. One feature of emerging pollutants is that while their presence in the environment is measurable, their consequences on human and environment health are often unknown (e.g. in case of degradation products over time) and their individual contribution to the overall health and environmental impact of the combined chemicals present in surface water is difficult to determine.

Such emerging pollutants may raise the future cost of treating water for drinking and before its discharge into the environment. The prevailing water management model in cities in OECD countries described in the opening section is inadequate to mitigate the risks generated by diffuse and emerging pollutants.

The resilience of freshwater ecosystems

Urban water security partly relies on ecosystems and the services they provide. Floodplains and wetlands can reduce the impact and occurrence of water-related disasters: they protect cities from floods and droughts by absorbing or storing excess water. The natural environment can also filter or dilute pollutants and improve the quality of the water accessed by cities. This is clearly illustrated by indirect potable reuse, where treated wastewater returns to water bodies (surface water or aquifers) before it can be used again: water bodies are used as environmental buffers, because natural systems have a high capacity to further purify water (Rodriguez, 2009). Surface water bodies also provide city dwellers with artistic, educational, recreational and spiritual well-being.

Cities often alter, degrade and reduce the resilience of water ecosystems. An estimated 25% of the world's rivers no longer reach the sea (Gaffney and Pharand-Deschenes, 2012) owing to a combination of malpractices, including by cities. The alteration of natural hydrological systems results in an increased run-off rate and volume; decreased infiltration and groundwater recharge base flow; deterioration of water quality in streams, rivers, and shallow groundwater; and loss of natural habitat and biodiversity. Climate change is also projected to cause adverse impacts on freshwater species diversity and water quality as a result of an increase in water temperature and changes in the physical, chemical and biological properties of lakes and rivers (Intergovernmental Panel on Climate Change, 2007).

Such degradation comes at a cost. McDonald and Shemie (2014) analysed 2 000 catchments serving over 500 cities worldwide. It calculated that large cities depend on an area the size of Russia to collect, filter and transport water. It further calculated that 700 million people could receive higher-quality water if cities fully invested in nature, and that the surveyed water utilities could save USD 890 million a year in water-treatment costs through watershed conservation.

Ageing infrastructure

Cities in OECD countries usually have adequate infrastructures to supply the water they need, collect and treat wastewater, protect them from floods and heavy rains, and alleviate the effects of drought. Access to safe water supply and coverage by sewerage networks (Figure 1.3) are distinctively good in OECD countries (OECD, 2012a).

Figure 1.3. **OECD population connected to wastewater treatment plants**

1990-2009, % of total population

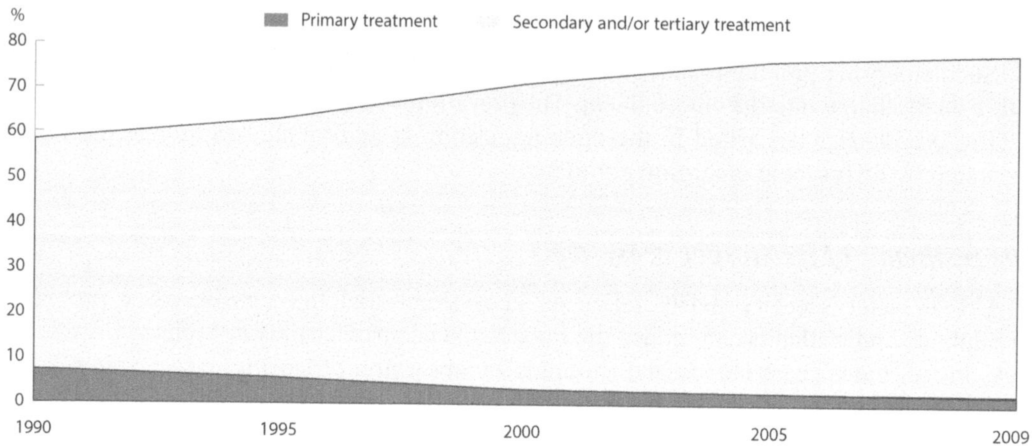

Source: OECD (2012a), *OECD Environmental Outlook to 2050: The Consequences of Inaction*, http://dx.doi.org/10.1787/9789264122246-en.

Recent developments have focused on the construction of wastewater treatment plants to improve the quality of wastewater discharge to the environment. European cities have adapted to the requirements of the European Union (EU) Urban Wastewater Directive 91/271/EEC established in the 1990s, but are lagging in renewal of existing infrastructure. While the status of WSS infrastructure is not known with accuracy, converging evidence in several countries suggests that urban water infrastructure requires significant investment to meet current and future challenges. Furthermore, networks are ageing and deteriorating; many water networks are nearing or past the end of their design life (Davis et al., 2013). For example, 75% of urban water networks in the United Kingdom are over 100 years old (Water UK, 2011); about half of the main water pipes in London are over 100 years old, and one-third could even be older than 150 years (London Assembly, 2003). In Los Angeles, 20% of the pipes were built before 1931 and nearly all will need to be replaced in the next 15 years, at an estimated cost of USD 1 billion; this will require doubling the current pace of replacement and tripling the related expenditures (Poston, 2015).

Ageing puts the efficiency of water infrastructure at risk. For instance, leaking pipes generate additional costs, both environmental (more freshwater is used and lost, and some wastewater returns to the environment untreated) and financial (through the opportunity cost of leakage and the cost of treating water that leaks before it reaches the consumer, thereby increasing the unit treatment cost). In the United Kingdom, the total water leakage has decreased from 30% to 22% during the last 20 years (Ofwat, 2012). While it is technically possible to further reduce the loss percentage, it is not currently economical (Beal et al., 2012). However, where demand for water increases and future supply potentially decreases with climate change, the future value of water will increase, and further improvement of infrastructure efficiency will become cost-effective (Karaca et al., 2014).

In addition to infrastructure inefficiencies and leakage, ageing infrastructure eventually leads to increases in functional asset failures and corresponding increases in risks of social and commercial disruption, as well as stronger impacts on human and environmental health (Davis et al., 2013). Failures such as water main bursts or sewage spills can also prove very costly to both the service provider and the water user (Davis et al., 2013); the risk and consequences of failure become greater with climate change, and the rising frequency and intensity of extreme atmospheric events coincident with the pressures of further urbanisation. Ageing infrastructures also generate future liabilities, as infrastructures decay. Hence, significant investments are required to replace or refurbish ageing and inefficient infrastructure for disaster preparedness as well as increase the structural resilience of water systems (Ray and Jain, 2014).

Finally, traditional urban water infrastructure in OECD countries is energy-intensive. Sanders and Webber (2013) note that universal access to a safe and secure water supply would be achievable in the absence of energy constraints. However, water requires energy for treatment and pumping: in the United States, 13% of annual primary energy consumption goes to water services (primarily water heating, but also water treatment). Both energy consumption and the cost of WSS services are increasing as clean, gravity-fed surface water sources become over-allocated and water is collected from increasingly distant, deep, or contaminated sources requiring large amounts of energy for pumping and treatment. Future water supplies may be more expensive and energy-intensive than the prevailing easy-access sources, which have been tapped first.

In Australia, Kenway (2014) calculated that urban water services indirectly utilise 8% of primary energy (i.e. 13% of electricity and 18% of natural gas). Heating and industry were the main users of energy, with utilities accounting for roughly 10% of energy consumption for urban water services. Energy costs are projected to rise due to increasing energy demand associated with socio-economic growth. Kenway (2014) notes that the energy demand for water in Australian cities will grow by 200-250% between 2007 and 2030. The energy bill for urban water management will multiply sixfold in most Australian states as a result.

The rest of this section reviews the gap in financing and general trends based on available data on water infrastructure needs and the related costs of urban water management. While international data on the subject are patchy and largely inconsistent, some conclusions emerge. They converge with selected country-level analyses that can be more robust, but difficult to compare across countries.

Global projections for financing needs for urban water infrastructure

Infrastructure needs for urban water management are difficult to measure. First, Lloyd Owen (2011) notes the assumption that WSS infrastructure assets need to be replaced after a fixed period: 40-80 years for water and wastewater distribution systems, 25 years for mechanical systems (pumping stations, as well as water and sewage treatment works) and 10-15 years for some sub-systems and components. In reality, asset replacement and refurbishment cycles can last considerably longer than the design life, resulting in a decline in the quality of the service and the value of the asset; in North London, Thames Water is facing 30-35% distribution losses from pipes laid 120-150 years ago. Second, there is no clear optimal performance level of performance for existing infrastructure. The author further notes that while there is no ideal leakage level, leakages above 15-20%

should be addressed as part of basic refurbishment and maintenance work to counteract contamination issues and loss-of-opportunity costs. In Germany and the Netherlands, leakage rates of 3-8% are considered close to the present realisable limit.

With this caveat, a strong consensus exists that global financing needs for urban water infrastructure are significant and increasing rapidly. Table 1.1 presents the results of three different studies on the global capital required for investing in water infrastructure (Box 1.4 provides more information). Global estimates ranged from USD 6.7 trillion by 2030 to USD 22.6 trillion by 2050. Needed investment in water infrastructure also varied markedly between the OECD (2006) and Lloyd Owen (2011) studies.

Despite these inconsistencies, attempts to project infrastructure needs for urban water management yield several important messages:

- In OECD countries, the water supply and wastewater system networks are generally the most valuable assets, comprising some 60-80% of the total value of all urban water and wastewater systems. In the United Kingdom, the current value of existing sewage assets alone is some USD 200 billion (OECD, 2006).
- Cities in OECD countries increasingly face critical future investment needs for water management to upgrade ageing systems and old technologies, as well as provide resilience against socio-economic and climatic changes.
- Investment needs for water management are large (significantly larger than for telecommunications, land transport, or electricity transmission and distribution), and finances are scarce (OECD, 2006).
- Although the benefits of investment likely outweigh the costs, it does not follow that these projected expenditures will be realised. Indeed, if past experience is any guide, such investment needs will likely not be met (OECD, 2006).

Country-level analyses provide additional detail on the status of urban water-management infrastructures, e.g. in France, Japan and the United States.

France

France has a large coverage of WSS infrastructures: the water supply networks span 850 000 km; the wastewater collection and treatment networks feature 97 000 km of combined sewer – 200 000 km for wastewater only and 95 000 km for stormwater collection. The estimated value of these assets is EUR 50 billion; the service quality can be assessed based on the indicators below (Office national de l'eau et des milieux aquatiques [ONEMA], 2012):

- compliance with biological quality standards:97%[3]
- compliance with chemical quality standards:98%
- service interruptions:[4] 4.43 per 1 000 users
- leakage rate for urban networks:[5] 21%.

Half of France's water supply networks were built before 1972 (Cador, 2002). The average network-renewal rate is 0.61 for water supply and 0.71 for wastewater collection and treatment, meaning that a full replacement of existing networks would take 160 years for water supply networks and 140 years for wastewater collection and treatment. The average renewal rate presents urban-rural disparities: services in high-density urban environments (≥200 habitants per km of network) have a significantly higher rate (1.31),

Box 1.4. **Projected investment needs for water infrastructure**

The three studies mentioned in this section build on different definitions and assumptions. They yield different results and present significant discrepancies, up to a factor of three between the Lloyd Owen (2011) and Booz Allen Hamilton (2007) projections.

Summary of projected investment required in global water infrastructure

USD trillion

	Investment period				
				2008-503	2008-50 [c]
Region	2005-15 [a]*	2005-25 [a]*	2005-30 [b]	Basic coverage	Full coverage
OECD	4.74	12.48		2.3	2.3
BRIC	2.98	8.28		2.2	2.8
OECD + BRIC	7.72	20.74		4.5	5.1
Rest of the world				2.2	2.9
GLOBAL			22.6**	6.7	8.0

a, b, c: See information on sources at the end of this box.

*Values are based on average annual estimates over this period. **This USD 22.6 trillion estimate for investment in global water infrastructure compares with USD 9 trillion for energy infrastructure, USD 7.8 trillion for road and rail infrastructure, USD 1.6 trillion for air and sea ports, and a total USD 41 trillion for urban infrastructures (Booz Allen Hamilton, 2007).

Note: BRIC = Brazil, Russian Federation, India, China.

Some studies focused on WSS, while others covered water infrastructure more generally. The Booz Allen Hamilton (2007) and Lloyd Owen (2011) studies were global, while the OECD (2006) study focused on OECD and BRIC countries only. Some studies only measured investment needs, while others included operation and maintenance costs. The Lloyd Owen study (2011) was geared towards developing countries and limited to new infrastructure development, while other studies factored in upgrading and renovation of existing systems. Projections by Lloyd Owen (2011) did not factor in climate change, the EU Water Framework Directive, new contaminants, or the presumption of widespread adoption of desalination and/or water reclamation.

Discrepancies in the studies also derive from the fact that some studies project investment needs, while others measure expenditures (which are likely to be lower than estimated needs). The treatment of refurbishment and replacement is also unclear: where it increases the serviceability levels of assets, it may qualify as capital spending; otherwise, it qualifies as maintenance. In any event, there is high uncertainty about water-asset refurbishment and replacement needs, particularly where infrastructure is buried.

While Booz Allen Hamilton (2007) does not disclose the basic data on which the calculations are based, the OECD (2006) and Lloyd Owen (2011) follow distinctive approaches and methodologies:

- The OECD report covers stormwater and urban drainage, and has a broader understanding of refurbishment. It is based on the assessment of past expenditures as a share of GDP at the country level. Projected levels were adjusted for a set of drivers that may affect investment needs (including innovation);
- The Lloyd Owen projections are based on the capital cost of water infrastructure to achieve a set level of coverage and service, as well as actual costs. Future costs did not incorporate innovation.

Sources: a: OECD (2006), *Infrastructure to 2030. Telecom, Land Transport, Water and Electricity*, OECD Publishing, Paris, http://dx.doi.org/10.1787/9789264023994-en; b: Booz Allen Hamilton (2007), "Lights! Water! Motion!", *Strategy+business magazine (Booz & Company)*, spring 2007, Issue 46; c: Lloyd Owen, D. (2011), "Infrastructure Needs for the Water Sector", unpublished paper commissioned by the OECD in preparation of the OECD report *Water Security for Better Lives*, OECD Publishing, Paris.

signalling faster renewal. Even so, 80 years would be necessary to renew existing water infrastructures in densely populated areas that may have exceeded the infrastructure's design life.

Japan

Japan has historically had high investment in water supply services. The country rapidly developed its water supply infrastructures after 1955, when it embarked on a period of high economic growth. The amended Water Supply Act prompted a second investment peak in the 1970s. A third peak occurred in the 1990s with the establishment of the "Plan for Fresh Water Supply" and subsidy for advanced treatment of water. Nevertheless, official estimates hold that significant future investment is required for water infrastructure in Japan (Ministry of Health, Labour and Welfare [MHLW], 2010), prompted by *(i)* the renewal of the ageing water infrastructure, most of which will need replacing in 20 years; and *(ii)* the need to strengthen the infrastructure to meet earthquake standards. These investment needs coincide with a projected decline in available financing, such that they will exceed the potential available funds for investment by 2025. Finding other sources of capital is therefore critical.

These projections rest on the following assumptions (MHLW, 2010):

- For future investment needs, the water supply facilities nearing the end of their design life will be reconstructed with the same function.
- Annual investment in water supply facilities will decline by 1% compared to the previous year. Since the investment capital is largely raised by subsidies from the MHLW, this assumption reflects the severe financial situation of the Japanese government.

Figure 1.4. **Investment needs in WSS in Japan**

trillion yen

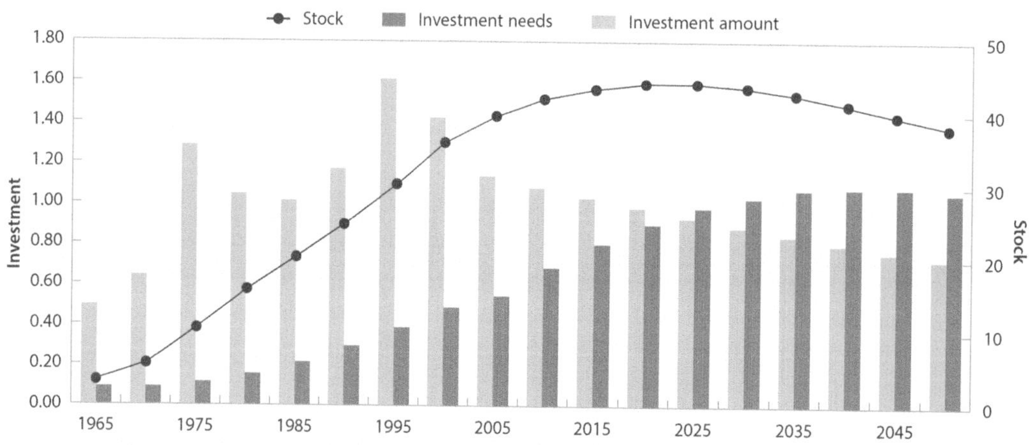

Source: MHLW (2010), www.mhlw.go.jp/shingi/2010/02/dl/s0202-8g.pdf (in Japanese).

The United States

The US Corps of Engineers plays a crucial role in ensuring the water security of US cities: it supplies water, supports flood risk management and regulates wetland alteration. To do so, it builds, operates and maintains extensive and diversified water-management infrastructures.

In a recent report, the US Corps of Engineers acknowledged that federal funds over the past 20 years have sufficed to maintain water infrastructure adequately (National Research Council, 2012). Similarly, local and state funds for water management have declined, with the result that the estimated value of the infrastructure managed by the US Corps of Engineers has dropped from USD 237 billion in the 1980s to USD 164 billion in 2010.

The funding gap over 2000-19 for clean watershed and drinking water capital, operation and maintenance costs is estimated at over USD 500 billion (USEPA, 2013b), excluding drainage costs. The investment capital required for the public water system infrastructure to continue to provide safe drinking water to the public over 2011-30 is estimated at USD 384.2 billion (USEPA, 2013b), excluding expenditures on raw-water dams and reservoirs, projects related to population growth, and the operation and maintenance costs of the water supply. These figures may need revising to account for climate change adaptation.

Acknowledging that the bulk of required infrastructure has been established, the National Research Council provides some recommendations on how the US Corps of Engineers could improve operation, maintenance and rehabilitation (OMR) of its existing water-related infrastructure, pointing towards alternative ways to enhance urban water security. The National Research Council (2012) considers that:

- Legislation (particularly the Water Resources Development Act) is geared towards building new infrastructure, and may be ill-adapted to the current issue of financing the operation and maintenance of existing assets.
- Opportunities exist for greater private-sector involvement in the US Corps of Engineers' OMR activities.
- Non-structural flood risk management options that "are more efficient, less costly, and provide greater environmental benefits" could be explored more systematically.
- Opportunities exist to collaborate further with local communities in order to lower costs and share the financial burden.

The USEPA (2013b) also suggests that raising public awareness of ageing infrastructure can increase support for the investment required to ensure access to safe and clean water.

Institutional changes affecting urban water governance

Respondents to the OECD survey on water governance in cities (OECD, forthcoming b) identified several drivers of changes in urban water governance (Figure 1.5). Some drivers (climate change and extreme weather events; water scarcity and competition; water pollution; and non-existent or ageing infrastructures), discussed above, are well established. Others (national and international laws and regulations; territorial reforms; decentralisation and reallocation of competences; and to a lesser extent, liberalisation and privatisation trends) are institutional reforms that deserve particular attention: even though they do not originate in the water sectors, they affect urban management.

Cities' governance structures affect the natural, capital and human resources available to manage water, the role of local authorities (regulators, facilitators and service providers) and the way cities deal with interdependencies across institutions, places and sectors. They can particularly foster co-ordination across levels of government, possibly reducing fragmentation. The international water-related political agenda and the implementation of the Human Right to Water and Sanitation[6] call for special attention to disadvantaged people, even in developed countries.[7]

Figure 1.5. **Key drivers affecting urban water governance**

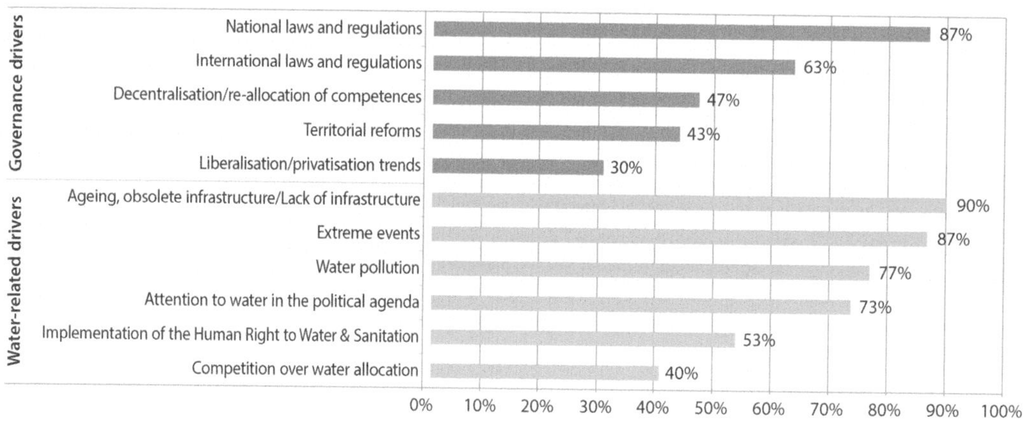

Note: Results based on a sample of 30 respondents who indicated the governance drivers as "critical" and "Important".

Source: OECD (forthcoming b), *Water Governance in Cities*, OECD Publishing, Paris.

This section discusses institutional drivers. It pays particular attention to national and international regulations, administrative and territorial reforms, and a new attention to stakeholders.

National and international regulations

Local governments and citizens play an active role in implementing national and international laws regulating several aspects of urban water governance. For example, the involvement of local actors and citizens in the European Water Framework Directive is key to managing water sustainably way, co-ordinating public action across government levels and reducing local conflicts. In most countries, national governments regulate or intervene in some local authorities' activities, monitoring and controlling their compliance with international human rights obligations and the relevant national legislation, regulations and policies (de Albuquerque, 2014). The recognition of access to drinking water and sanitation as a basic human right (United Nations Resolution 64/292 of 28 July 2010) stimulated initiatives, such as the "Water, a Human Right" European Citizens' Initiative, which collected 1.9 million signatures across Europe.

Two additional items serve as illustrations. The first item is environment and health standards, which drive the costs of water security and water services. The European Union Directive on urban wastewater treatment 91/271/CEE identifies a total of 72.8% of EU28 territory as sensitive areas/catchments that require more-stringent wastewater treatment before discharge to water bodies in order to reduce the risk of eutrophication.[8] The catchment surface area that is considered sensitive increased by 4.8% between 2011 and 2013 (European Commission, 2011). The Directive also establishes an obligatory timetable for providing improved urban wastewater collection and treatment systems. A 2011 review indicated that more-stringent treatment installations were already in place for 77.3% of the total generated wastewater load of big cities. The Swiss Federal Office for the Environment (FOEN) recently mandated that the largest 100 wastewater treatment plants in the country add a fourth step to the treatment process, thereby reducing by half the amount of micro-pollutants in Swiss rivers and lakes. The estimated cost of this additional water treatment is CHE 1.2 billion (Swiss francs) (FOEN, 2012).

The second item is technical standards. Indeed, urban water management in selected countries may be biased towards high-cost options. Lloyd Owen (2011) notes that Austria and Germany have adopted a high-cost, arguably over-engineered approach towards WSS service provision, driven by national and EU-related imperatives. Another driver of urban water management is industrial policy to support domestic the engineering and construction sectors. Over 1990-2010, England and Wales financed a major catching-up exercise to remedy under-spending in the 15 previous years and accepted previously contested EU directives (Lloyd Owen, 2011). During the period, England and Wales spent GBP 90 million (pounds) renewing a dilapidated network, while Germany spent EUR 110 million upgrading a network (in the former West Germany) considered to be in materially better condition (Lloyd Owen, 2011).

Administrative and territorial reforms

Administrative and territorial reforms across OECD countries are radically changing the municipal landscape. In 2007, Denmark reduced the number of municipalities from 271 to 98; Turkey reduced the number of municipalities from 3 225 to 2 950 in 2008, and further decreases (from 2 950 to 1 395) as of March 2014. In 2007, Finland started restructuring local government and services (the "PARAS reform"), encouraging voluntary municipal mergers and co-operation for public service delivery (OECD, 2013a). In the Netherlands, the number of municipalities is more than halved following several mergers and re-organisations in the last 6 decades, and ongoing discussions target a threshold of 100 000 inhabitants per municipality (OECD, 2014). Other territorial reforms have taken place in Australia, Canada, Germany, Iceland, Norway, Switzerland and the United Kingdom.

The current economic crisis and tight budget policies have given such reforms further impetus, and municipal-merger policies have picked up as austerity measures take their toll. The economic and financial crisis has provided many governments with an opportunity to step up municipal re-organisation, with the goal of rationalising and pooling resources to increase the efficiency of local public action.

In parallel to these territorial reforms, a number of OECD countries are also experiencing deep reforms aiming to consolidate and corporatise water operators. Italy numbered 3 704 water service operators and 4 278 wastewater service operators (Marques, 2010) before the 1994 and 2006 reforms. Since then, consolidation has taken place. Some 2 000 operators were active in the water sector in 2014, and their number is further expected to shrink to 91. Meanwhile, the consolidation of the Irish water industry is very recent or still ongoing (Box 1.5).

Box 1.5. **Consolidation of the water industry in Ireland**

Irish Water is Ireland's new national water utility, responsible for providing and developing water services throughout Ireland. Incorporated in July 2013 as a semi-state company under the Water Services Act 2013, Irish Water will bring the water and wastewater services of the 34 local authorities together under one national service provider, taking over their responsibilities on a phased basis since January 2014. Irish Water will require approximately five years to be fully established, at which point it will be responsible for operating all public water services, including managing national water assets, maintaining the water system, investment and planning, managing capital projects, and customer care and billing. Irish Water will also be making capital and investment decisions regarding the country's water infrastructure on a national basis. It will answer to two regulatory bodies – the Commission for Energy Regulation, which is the economic regulator for the water industry, and the Environmental Protection Agency, which is the environmental regulator.

Source: OECD (forthcoming c), *The Governance of Water Regulators*, based on www.water.ie/about-us/company/about-irish-water/.

A new attention to stakeholder engagement

Stakeholder engagement is expected to play a larger role in urban water management and hence deserves particular attention. It will likely require innovative governance arrangements to consult with a variety of stakeholders and leverage their capacity to manage water.

OECD work on water security argues that setting an appropriate level of security is not merely a technical exercise: perceptions matter, and can best be considered through stakeholder consultation (OECD, 2013b). OECD city dwellers may be surprised by the fact that their water bill does not reflect their decreased water consumption (when conservation policies increase the unit costs of water to stabilise revenues for water utilities) (Chapter 2). Innovation in urban water management often relies on actions taken by water users, e.g. conservation practices and investment in water-saving appliances (Chapter 3). It follows that stakeholder engagement is likely to play an increasingly critical part in urban water governance.

Overcoming current and future water-related challenges in cities in OECD countries will require effective governance frameworks that pay particular attention to stakeholder engagement. Chapter 5 focuses on selected dimensions, building on recent developments of the OECD Water Governance Initiative.

A framework for urban water management

This section sketches an analytical framework for achieving a level of water security for cities in OECD countries at the least cost to society. The framework brings together four interlinked dimensions cities can explore to enhance their water security now and in the future. One dimension is financing, as cities in OECD countries rely on extensive infrastructures that need renewing. Another dimension is innovation (both technical and non-technical), which can lower the costs of water security. A third dimension is the urban-rural interface: some innovations, particularly green infrastructures or ecosystem services, require cities to co-operate with their surroundings. Finally, institutional arrangements help combine the different scales at which cities manage water, regulate urban water services and engage with stakeholders – who have their say in the level of water security they deem appropriate, how much they are willing to pay for it, etc.

- Financing: what are the options to lower the cost of responses to water challenges (e.g. through enhanced asset management or exploiting low-cost water sources)? How can traditional financing sources (particularly tariffs and taxes) best be mobilised? What are the options for harnessing additional sources of finance (e.g. new sources of revenues for water utilities or private capital)?
- Innovation: which technical (e.g. smart water systems and distributed systems) and non-technical (e.g. water-sensitive urban design) innovations can contribute to water security and affordability in cities in OECD countries? What are the barriers to their diffusion in cities in OECD countries, and how they can be overcome?
- The urban-rural interface: what are the benefits of improved co-operation between cities and their rural surroundings? How can such co-operation be put into practice?
- Governance: how can urban water governance meet future challenges? Particular attention will be paid to the combination of different scales for urban water management, stakeholder engagement and the regulation of WSS services.

Figure 1.6. **A framework for city-level water management**

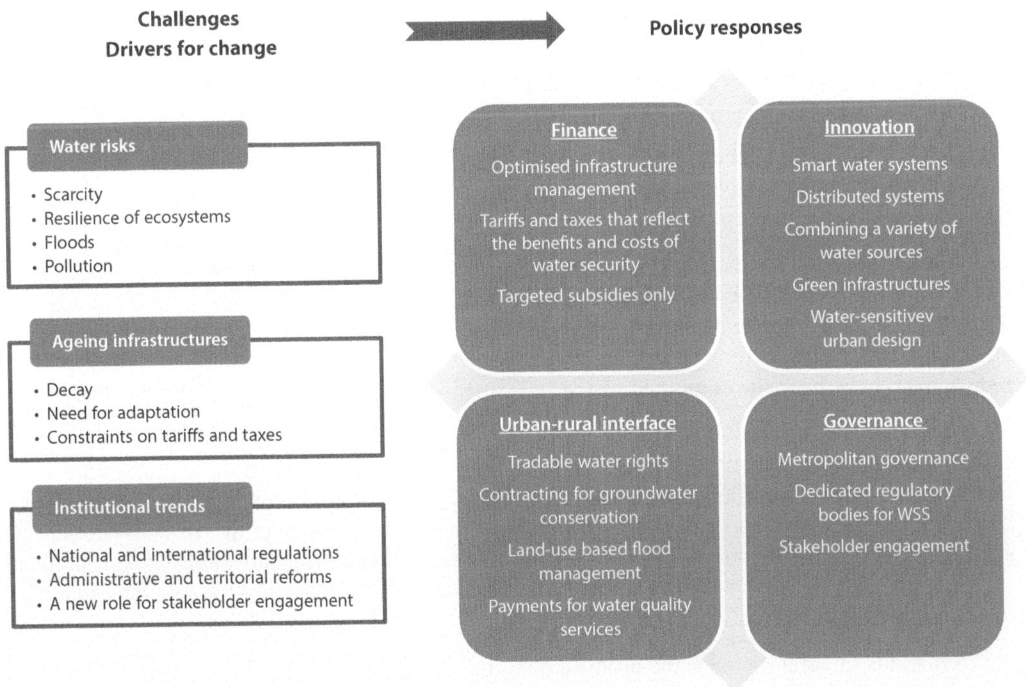

The case study of Auckland (Box 1.6) illustrates the complex interactions among the four dimensions. It shows a most interesting combination of measures, dealing with financing and business models (harnessing property developers and offsetting), technical innovation (stormwater management and multipurpose green infrastructures), non-technical innovation (water-sensitive urban design, catchment and aquifer accounting), scale (amalgamation and urban-rural co-operation) and governance (stakeholder engagement and long-term vision reflected in spatial planning).

The following section sketches three sets of scenarios for future urban water management in OECD countries. It illustrates how distinct combinations of the four dimensions mentioned above lead to alternative futures.

Box 1.6. A multifaceted approach to urban water management: Auckland, New Zealand

Auckland's sub-tropical climate – it receives an average of 1 240 mm of rainfall a year – and location between two sensitive and highly valued harbours make stormwater management a core feature of the Auckland Council's role.

Auckland's most pressing water management-related issues derive from its rapidly growing population and expanding urban footprint. Wherever Auckland has expanded into greenfield areas, freshwater and highly valued marine receiving environments have been degraded. The new requirement to manage within nationally prescribed bottom lines for freshwater quality elevates the importance of freshwater management as a driver of urban form and design in Auckland, and makes it a key factor in the Council and government's plans for managing growth and meeting urban growth demands.

In 2010, Auckland reformed its governance arrangements, with the goal of delivering an agreed regional vision and build the economies of scale required to deliver it. The region's seven district councils (responsible for managing land and delivering council services) and the Auckland regional council (responsible for managing the environment and regional strategy) were amalgamated to create a single unitary authority.

Amalgamation served to centralise the council's research, investigation and monitoring functions. It has allowed the council to develop a centre of excellence for science, generated the scale necessary to complete high-quality work and ensured that policy advice and political decision-making are underpinned by the best evidence available.

It also gave the Council and Council-controlled organisations the necessary scale to tackle issues previously beyond the reach of individual councils. The Council has been able, for instance, to upgrade substantially the region's two key wastewater treatment plants and advance the NZL 950 million (New Zealand dollars) "central interceptor" project, which will reduce overflows from the combined waste and stormwater system of the Auckland isthmus.

The Council is currently initiating a "Greenways" programme to align Council actions and investment across a range of policy and operational units, with the aim of delivering multiple freshwater, biodiversity, transport, urban design and stormwater-related outcomes from the same investment. To support this, it is developing the necessary capability to convene and support cross-council "virtual teams" bringing together experts from different backgrounds with community and stakeholder groups to tackle common issues and promote integrated, location-specific outcomes.

While this integration holds the promise of significantly improved and more cost-effective outcomes, the Council recognises that this action alone will likely not suffice to meet statutory bottom lines for freshwater quality, especially in already-degraded urban catchments. By maximising the integration of council planning, programmes and investment, and catalysing collaboration with community and private-sector providers, the Council hopes to catalyse innovative responses to complex and emergent issues pertaining to water quality.

Source: the case study was developed by Andrew Schollum and Grant Barnes, Auckland Council, for this OECD project.

Financing

As argued above, cities in OECD countries need to upgrade, renew and possibly adjust the infrastructures they use to manage water risks and deliver water services. Financing is therefore an essential feature of future water management.

A number of countries and cities have postponed infrastructure renewal for several decades. Their asset-management policies rested on the premise of cheap water, reflecting neither the opportunity cost of the resource nor the cost of upgrading existing infrastructures. A recent assessment of WSS cost recovery through pricing in Europe suggests that water companies either have relied on hidden subsidies for necessary capital investments or still rely on the good status of assets built with public funds (European Environment Agency, 2013).

Underinvestment in infrastructure has resulted in declining asset value and increased risk of functional failures, social and commercial disruption, as well as negative impacts on human and environmental health. While further postponing investment may generate higher damages or transition costs, it could also minimise the risk of inadequate infrastructure adaptation to changes in water availability and demand (by reducing uncertainties related to climate change and population dynamics), as well as provide access to new and more efficient technologies or practices. Timing, and the capacity to avoid lock-in in specific technical trajectories, will be essential.

An appropriate combination of revenues from water tariffs, taxes and transfers (the 3Ts) from the international community is critical to providing sustainable water services (OECD, 2010). Additional sources may be available to bridge a financing gap, but will be repaid through a combination of the 3Ts. Since each of these sources is – and is likely to remain – severely constrained, financing urban water management in OECD countries will increasingly require efforts to: *(i)* minimise costs for urban water management through optimising maintenance and exploiting low-cost water sources; *(ii)* design water tariffs to secure stable revenues while serving water policy objectives and addressing equity issues; *(iii)* ensure that those benefiting from water security or increasing its costs also contribute financially; *(iv)* access private equity. Chapter 2 discusses the role of financing in achieving water security, including providing WSS services.

Innovation

Innovation can minimise the operating costs of water management and (to a certain extent) postpone renewal. Cities in OECD countries would benefit from two sets of innovations:

- Technical innovation can enhance water security, e.g. refined-membrane technologies can treat water and wastewater to higher-quality grades. Investment will be required to fit such innovative tools into existing networks. Other innovative technologies can increase efficiency and save costs, e.g. methane can be captured in wastewater treatment plants and utilised as a source of energy, thereby minimising energy costs.
- Non-technical innovation matters, e.g. innovative business models for water utilities and water-sensitive urban design, can minimise the cost of water security and prevent future liabilities. Contractual arrangements between municipalities and water users can share risks related to water management and the financial burden of achieving water security.

Green infrastructures deserve particular attention. They provide ecosystem service benefits and are an emerging planning and design concept to mitigate urban water risks. A range of green infrastructure options, often more cost-effective than traditional engineered approaches, exist to enhance water security. City managers will need to address trade-offs between alternative techniques featuring different combinations of water, energy and land.

Innovation takes place either in the context of new greenfield developments[9] (as cities expand or build new areas or facilities, such as a hospital or airport) or brownfield projects. Deploying innovative techniques and practices in OECD countries will often require retrofitting existing infrastructures – which introduces particular challenges worthy of attention – and reforming prevailing practices. The diffusion of innovative water management in cities in OECD countries requires water prices reflecting the resource scarcity and externalities associated with water security, as well as institutional and regulatory frameworks supporting sustainable forward-looking urban water management practices (see Chapter 3 for a more detailed discussion).

The urban-rural interface

Co-operation between cities and their rural surroundings allows for more effective water management at the river basin scale, as both the drivers of the problems and the consequences of the chosen policies have significant effects across different areas, jurisdictions and economic sectors.

Urban-rural co-operation can help cities enhance water security (see Chapter 4). For example, compensating farmers when their land is flooded can be more cost-effective than building dykes, which are less flexible and may become ill-adapted to future risks. Similarly, compensating upstream farmers for attaining quality objectives (by reducing the use of fertilisers, pesticides and herbicides) can be more cost-effective than treating polluted waters downstream for potable use.[10] Co-operation with other water users in the river basin can also mitigate competition for access to limited water resources, particularly during droughts. Cities can also contribute to enhanced water security in their environment by minimising sealed surfaces and utilising water-sensitive urban design to recharge aquifers, as well as limit groundwater and surface water contamination.

Potential synergies between urban and rural areas will only materialise when appropriate incentives and tailored arrangements are in place. What does it take for farmers to embark on sustainable agricultural practices to protect a catchment, or to accept their property be flooded to protect the city downstream? How can national and local governments create the appropriate collaboration and outcomes? Some tailored arrangements can translate into contracts across urban cores and rural hinterlands.

Water governance

The management of urban water presents multi-level governance gaps, some of which are compounded by the emerging challenges (e.g. administrative, policy or funding gaps) discussed above. Others stem from the change in the "governance climate" and the call for more inclusive, bottom-up policy making (e.g. accountability, information and capacity gaps). Three issues deserve particular attention:

- How can the different water management scales – from buildings (for decentralised systems) to catchment (when cities co-operate with their surroundings) – be reconciled?
- How can urban WSS services best be regulated?
- How can stakeholders contribute to the design and implementation of effective urban water management?

Chapter 5 considers future developments in urban water governance, considering a range of tools – e.g. metropolitan governance, dedicated regulatory bodies and institutionalised or inclusive forms of stakeholder engagement – that can help bridge governance gaps and support more sustainable urban water governance.

Alternative scenarios for urban water management

Three sets of scenarios for urban water management help illustrate how the four dimensions outlined above (financing, innovation, the urban-rural interface, and governance) can be combined, prefiguring alternative modes for managing urban water to achieve distinctive economic and environmental performance, as well as social equity. The scenarios were developed in the context of the Eau&3E research project (see Barraqué and Isnard [2013] for a synthesis). While the project focused on France, most of the lessons learnt are relevant to OECD countries.

A first set of scenarios stays within the limits of the prevailing model for urban water management: revenues from water bills remain the main source of financing; technological trajectories, and environmental and health regulations, remain basically unchanged. This situation will likely lead to a rapid decline in the quality of the water supply and treated wastewater, additional pressure on water resources and severe affordability issues. Some adjustments could take place: on tariffs, to reflect social concerns and incentivise water efficiency, and on infrastructures, to rely on pipeless technologies when networks have reached their limit. However, these adjustments do not translate into more sustainable urban water management. A more aggressive attempt to control demand is an option, but may trigger resistance from water users and generate dire financial consequences.

A second set of scenarios leads to a radical revamping of the prevailing model, in which the public good dimension of selected elements in the water cycle is revised, alternative sources of financing are explored, and taxpayers finance wastewater treatment. This new model could also consider cross-subsidisation of water services or mutualisation of some support functions, leading to a re-organisation of water services. It could question the existence of a unitary service, exploring instead a dual water treatment that provides basic universal services while parts of the city access improved services at a higher cost. While these options could contribute to financial sustainability, they raise other concerns, including issues of political acceptability and equity.

A third set of scenarios recombines the management scale of water services, managing urban water at a larger scale to mutualise access, benefit from economies of scale and provide territorial solidarity. The risk is that urban water systems may reach diseconomies of scale as they supply distant city dwellers. To remedy this, they can be managed at smaller scales, combining both piped and non-piped technologies where appropriate. Such approaches can be driven by central states, leading to territorial re-organisation. Alternatively, they can stem from combined initiatives by urban and rural communities – such as payment for ecosystems services, whereby downstream cities pay upstream farmers to preserve natural ecosystems in order to secure water quantity and quality. Cities can also rely on their own devices, in isolation from their environment, to secure the necessary water through desalination and systematic water reuse.

The three sets of scenarios highlight that urban water management can evolve in different directions. They reflect social preferences in terms of the appropriate level of water security, solidarity across territories and user groups, or the relative values attached to water, land or energy. While such preferences may be implicit, the scenarios help anticipate some of their consequences and prefigure the cities and societies we will live in.

Clustering cities and policy responses to water management

This section proposes a working definition of cities based on previous OECD work. Administrative boundaries alone cannot account for the challenges facing cities; other criteria are relevant to approaching urban issues. The section outlines a typology to cluster cities as regards water management.

A working definition of cities

The OECD has traditionally used thresholds based on population density (the ratio between population and the total area of the administrative unit) to classify regions as either urban or rural. While this approach has the obvious benefit of simplicity and performs well for several applications, it has clear limitations when applied to the analysis of urbanisation patterns and their effects on the economy, the environment and social relations.

While the concentration of people in dense urban centres of "established" cities in OECD countries has slowed down or even decreased in some cases, other agglomerations of varying sizes (including London, Milan, Tokyo, Manchester and Lyon) have continued to grow, sometimes through their geographic footprint rather than increased population density. Some urban areas are evolving from monocentric agglomerations to more complex systems comprising integrated urban centres (cores) and sub-centres. In other areas, a number of cities and towns are increasingly linking up, forming polycentric integrated areas. This changing spatial organisation, and the wider territories within which cities are located, directly affect the quality of life of their inhabitants, the demand for transport infrastructures, the surrounding landscape, the directions of human and capital flows, and the global environmental footprint of urbanisation.

The OECD (2012b) notes that monitoring urbanisation and comparing the performance of urban areas require new definitions based on economic function rather than administrative boundaries. A definition of urban areas as functional economic units can better reflect these changes and guide the way city governments plan infrastructure and services. The concept acknowledges:

- the growing consensus that public policies should be concerned not only with the scale, but also with the geography of urbanisation, and that the functioning and efficiency of linkages between cities, and between urban and rural areas, can lead to important changes in how and where economic production takes place
- the role of large metropolitan areas in the global economy, and their capacity to realise the benefits of economic agglomeration, industrial clustering and innovation
- the potential of medium-sized cities to drive more sustainable urban development, without the costs and inefficiencies associated with mega-cities.

In collaboration with the European Union (Eurostat and EC-DG Regio), the OECD (2012b) has developed a harmonised definition of urban areas as "functional economic units", thus overcoming previous limitations linked to administrative boundaries. The functional approach better fits the dynamism of urban contexts, using population density to identify urban cores and travel-to-work flows to identify the hinterlands whose labour market is highly integrated with the cores (OECD, 2012b). It has helped identify, for each OECD country, all urban systems with a population of at least 50 000. Chapter 5 discusses institutional arrangements to manage water in cities.

A typology of cities for water management

Cities are exposed to and affected by water challenges in different ways, and their capacity to respond also varies. The section below outlines a typology of cities in OECD countries relevant to water management combining two sets of criteria: a) the challenges cities face with regard to water management; b) their capacity to respond to these challenges, depending on their exposure to water risks, their institutional architecture and the features of the urban environment. Cities can use these criteria to situate themselves among similar cities and tailor their responses to future water challenges.

Similar attempts found in the literature pave the way. Annex 1.A1 captures relevant groupings developed by Fernandez (2014) and van der Steen (2011).

The typology outlined below draws on considerations developed in this chapter, as well as OECD work on urban development and territorial indicators. It combines three dimensions:

- exposure to water risks: floods, scarcity, pollution and ecosystem resilience;
- distinctive urban features: affluence, energy endowment, surroundings, size, urban dynamics, spatial patterns; and
- institutional architecture.

The rest of the section explains in more detail the different dimensions (Figure 1.7) and the criteria designed to capture them.

Figure 1.7. **Criteria to cluster cities as regards water management**

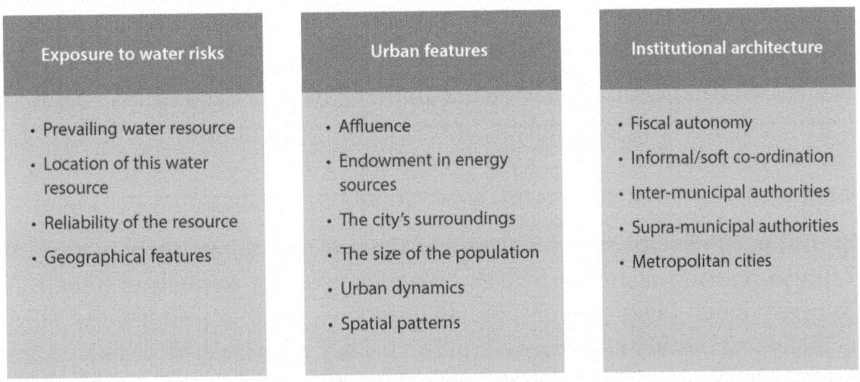

Water risks

This chapter has discussed water risks in detail. Relevant criteria to characterise a city's exposure to such risks include: **the prevailing water resource** (surface versus groundwater); **the location of the resource** (local versus distant – distant sources require infrastructure to transport water and are more likely to engender conflicts with other distant users); and **the reliability of the resource** (renewable versus non-renewable source; level of water stress). Geographical constraints also matter. For instance, physical relief has very significant impacts on urban run-off, as well as on the cost of installing and operating water and wastewater systems.

Urban features

Several urban features affect a city's exposure to water risks and its capacity to respond:

- **Affluence** drives city dwellers' water consumption and demand (in terms of both quantity and quality), as well as infrastructure design and financing capacities. It also affects the value of the assets at risk. A simple criterion to measure affluence is whether GDP per capita is above or below the average for OECD urban areas.
- **Endowment in energy sources** affects construction and operation costs, and can drive trade-offs between water, energy and land. While it may be difficult to identify a simple metric, Fernandez (2014) provides preliminary data.
- **Urban surroundings** can be a source of environmental risks (e.g. in coastal zones) and vulnerability to human impact, but can also provide ecosystem services. Appropriate categories are urban, rural, coastal zones and deltas.
- **Population size** affects exposure to risks and raises issues of scale. Large cities have more opportunities to collect revenues and invest in water security; they accumulate expertise and technical capacities to plan, manage and operate water infrastructures. Small cities face lower competition with water users, discharge less wastewater run-off and have less impact on the resilience of freshwater ecosystems. In Chapter 5, cities are clustered in the following categories: *(i)* below 500 000 inhabitants; *(ii)* between 500 000 and 1.5 million inhabitants; *(iii)* between 1.5 million and 5 million inhabitants; and *(iv)* above 5 million inhabitants.
- **Urban dynamics** generate both risks and opportunities. Growing cities have opportunities to invest in greenfield developments and new infrastructures, and raise additional revenues (including local taxes). However, they need to keep pace with population growth when planning and investing in water security and water services. Shrinking cities may be encumbered by oversized infrastructure and fail to generate the revenue required to operate and maintain them. Urban dynamics can be measured by the city's population growth rate compared to average population growth in cities (OECD Metropolitan Database[11]).
- **Spatial patterns** affect water risks and options to mitigate them. Cities in OECD countries can be distinguished as either compact or sprawling (OECD, 2012c). Compact cities seem better able to conserve land resources for agriculture, recreation, and water and energy provision. They can also save costs to connect city dwellers to water services. However, compactness can make retrofitting existing urban water infrastructures more costly and affects the water-energy-land trade-offs: water management options that are less energy-intensive but require more space may be inappropriate in compact cities. Urban sprawl can affect ecosystems when natural habitats are destroyed or fragmented by developed land. A relevant criterion is the sprawl index, which enables classifying cities according to their measured growth in built-up areas, adjusted for the growth in city population[12] (see Chapter 5 for its application to selected cities).

Cities by institutional architecture

Cities' institutional architecture can affect urban water management in several ways. Depending on how responsibilities are shared, it can facilitate (or hinder) co-operation at the appropriate geographical scale, co-operation across policy domains and autonomous decision-making on issues of water security. Metropolitan and hydrological boundaries rarely match; aligning views, interests and motivations must take into accounts all the stakeholders in the basin. Similarly, metropolitan boundaries can encompass several municipalities, even in the absence of a metropolitan authority (OECD, 2011b).

A first measure of a city's capacity to take initiative and respond to water challenges is **fiscal autonomy**, i.e. its capacity to raise revenues to implement its policies. This capacity dictates whether the city has some capacity to adjust revenue streams to needs or instead depends on fiscal transfers from a central budget. It can also serve as a proxy for political autonomy.

OECD work on urban governance highlights four types of institutional architecture relevant to water governance (see Kim, Schumann and Ahrend, forthcoming). Chapter 5 explores in more detail how these four groups relate to water management:

- **informal/soft co-ordination**, which takes the form of lightly institutionalised platforms for information-sharing and consultation
- **inter-municipal authorities**, which share costs and responsibilities across member municipalities
- **supra-municipal authorities**, an additional layer above municipalities, which can result either in a directly-elected metropolitan government or in upper governments establishing a non-elected metropolitan structure
- **metropolitan cities**, a special status for largely populated cities, which puts them on the same footing as the next upper level of government and gives them broader competences.

The typology described above can serve to tailor analyses and policies to specific situations. The next section highlights potential matches between water risks and related options.

Tailoring responses to risks

Cities in OECD countries will respond to water management challenges by designing a long-term vision for urban development that guides the design and development of projects at any scale, e.g. building new facilities (hospital or transport infrastructure) or reconstructing buildings and districts. These responses will stimulate initiatives from a variety of local entrepreneurs and stakeholders. They will address equity issues by analysing the distributional effects of water challenges and alternative options, and implementing targeted social measures.

Some of the options presented in Table 1.1 and discussed in the following chapters target specific water risks. The table does not mention the risk to the resilience of water systems. As Chapter 3 argues, this risk is essentially addressed through green infrastructures, and any measure promoting them will help mitigate it.

The table also does not mention governance because, as argued in previous OECD work on water security (OECD, 2013b), governance arrangements need to be adjusted to risk severity. Water policy dialogues in the Netherlands (OECD, 2014) and Brazil (OECD, forthcoming d) again emphasised this point.

Table 1.1. **A menu of options tailored to specific water risks**

Abundant water	Scarce water	Polluted water
Stormwater management – i.e. porous pavement and green roofing	Water conservation measures in urban and agricultural areas	Stormwater management – i.e. porous pavement and green roofing
Taxes on impervious surfaces	Groundwater conservation policies	Taxes on impervious surfaces
Payment for ecosystem services (PES) to restore natural grasslands or wetlands for flood protection	Smart water systems and water distribution control	PES or contractual arrangements to improve water quality
Land use-based flood protection	Distributed water systems, where they facilitate water reuse in the city	Groundwater conservation policies
Dynamic network management and smart control	Water prices that signal scarcity	Retrofit existing stormwater ponds to include quality control
	Tariff structures that secure revenues when the volumes of water sold diminish	Pollution-prevention and pollution-discharge regulation
	Beyond-the-meter services, such as alarms to warn customers of leaks	Dynamic treatment optimisation and smart control
	Command-and-control instruments: technology standards, rationing policies, priority allocation	Use of natural pollution filters (e.g. natural or constructed wetlands)
	Reuse and recycling of water	

Beyond cities: A role for other tiers of government

While leaving wide latitude for local authorities to shape policies that fit local contexts, higher levels of governments can take a variety of initiatives to promote urban water management. Table 1.2 lists some of these initiatives, clustered around regulation,

Table 1.2. **Initiatives by other tiers of government to promote urban water management**

Type of action	Illustrations
Regulation	Establish codes for land-use zoning (especially in hotspots), spatial planning (e.g. mandatory water tests, as in the Netherlands), and building and construction (e.g. technical standards for sustainable drainage systems, developed in Scotland by the environmental regulator for housing developments). Develop pollution control regulation.
	Promote regulation on reclaimed water, clarifying where reclaimed water is beneficial and how it should be factored in water management plans.
	Set public procurement rules reflecting long-term perspectives on water and value flexibility in water management.
Resource provision	Provide resources to collect and share data and information across municipalities and different levels of government (e.g. identifying, sharing and applying good practices, scaling up successful experimentation and linking to sources of expertise).
	Facilitate city-to-city peer learning (e.g. by promoting networks such as "100 Resilient Cities").
	Provide a repository of information to allow comparing information across cities. Water regulators can provide a platform to collect and disclose information related to regulation and the performance of water services in their area of expertise, as well as contribute to developing common methodologies to compute performance indicators and other information.
	Ensure capacity development (including training and motivation).
Incentives	Use infrastructure financing, where available, to incentivise water-sensitive urban development or reward water-sensitive projects.
	Provide awards for urban water management.
	Devise water allocation mechanisms that adjust to shifting circumstances and reflect the distinctive capacity of cities, farmers and other users to mitigate water risks.
	Encourage mechanisms at the metropolitan level to pool resources and capacities across municipalities, and between urban and non-urban areas. A range of mechanisms apply, from informal co-ordination to metropolitan bodies.
	Raise awareness and enhance engagement of people and institutions that have a stake in the outcome or are likely to be affected.

resource (e.g. information and education) provision and incentives. Box 1.7 provides some illustrations of co-operation between tiers of governments on urban water management.

Box 1.7. Co-operation on water management between cities and other tiers of government

Israel

Two-thirds of the territory of Israel is water-scarce. Solutions for managing urban water take into account the national water frameworks.

The national water authority of Israel has set a 2050 strategic plan for water and sanitation. In 2001, municipalities were asked to create professional, financially autonomous water and sewage-management agencies comprising representatives from local governments. These agencies have contributed to improving infrastructure, with 30% of revenues from water and sewage collection invested back into repairs and maintenance. They have also managed to reduce water leakages.

Future urban water management will rely on innovative solutions. Among others initiatives, the Ministry of National Infrastructures, Energy and Water Resources launched a project on "Water in Smart Cities". Smart-city systems aim to address urban and rural water challenges (i.e. related to irrigation) and support the purification of industrial and domestic waters.

Israel also adopted a National Strategic Plan to recycle and reuse water for agriculture irrigation (50% of water used for agriculture is now recycled); 50 water recycling centres had been established by the end of 2014.

Scotland

Scotland has a largely centralised governance framework for urban waters featuring geographically harmonised water charges.

Scottish cities are faced with ageing infrastructures. Glasgow, for example, undertook a ten-year project to renew its wastewater system, partnering with the central government to better understand local realities and gather extensive data (e.g. on water and wastewater flows modelling). Knowledge creation and innovation helped bring the cost of the project down to GBR 600 million (pounds). The city's green infrastructure agenda, as well as its strategies on flood protection and wastewater-flow management, also fostered synergies.

Lessons learnt include the need to maximise co-benefits through partnerships between central and municipal authorities, engage with all stakeholders involved and encourage synergies with the long-term vision for the city.

The United States

The USEPA has ten regional offices working at city-level to support communities in improving infrastructure, as well as drive innovation and technology through grant allocation. The EPA also partners with municipalities to develop data on water quality and infrastructure. It provides them with economic incentives, such as revolving funds for clean water and drinking water. The funds are allocated to the states and go into developing municipal bonds to support various infrastructure needs.

The EPA has also engaged in urban water federal partnerships with 13 federal agencies and cities to connect and reconnect urban catchments and rivers degraded by industrial activities. In 2014, it awarded USD 2.1 million to 46 organisations in selected US cities and Puerto Rico.

Box 1.7. **Co-operation on water management between cities and other tiers of government** *(continued)*

The United States has also developed "water clusters" where universities and corporations can simultaneously develop and support new technologies. Ten water clusters are now in operation.

At the federal level, the government also provides green-building certification acknowledging water conservation, irrigation and energy production through wastewater. The federal government also produced guidelines in 2012 to set minimum quality standards for the different qualities and uses of reclaimed water.

Source: contributions from OECD delegates at the 4th meeting of the OECD Water Governance Initiative.

Notes

1. In this report, water security is defined as maintaining acceptable levels for four water risks: risk of shortage (including droughts); risk of inadequate quality; risk of excess (including floods); risk of undermining the resilience of freshwater systems. See OECD (2013b), for a detailed discussion of water risks and policy responses.

2. Water scarcity can have cascading effects, such as forest fires. Greece, for instance, is projected to face serious problems with droughts and forest fires, with floods, soil erosion and desertification as secondary effects. As noted in Chapter 4, forest conservation (and protection against fires) can shield cities from both floods and water scarcity by absorbing or storing excess water and regulating the hydrological cycle.

3. Another source (Direction Générale de la Santé, système SISE-Eaux) indicates that 95.2% of France's population receives with water that complies with biological standards more than 95% of the time (2003 data; compare with 91.5% of the population in 1999).

4. Defined as the number of interruptions in water supply for which water users had not received advance notice.

5. It is noteworthy that the leakage rate diminishes in regions where the main source is surface water; however, surface water often needs costly treatment before it can be used for domestic consumption.

6. On 28 July 2010, the General Assembly of the United Nations adopted the Resolution A/RES/64/292, which "recognises the right to safe and clean drinking water and sanitation as a human right that is essential for the full enjoyment of life and all human rights". The General Assembly adopted the Resolution by recorded vote of 122 in favour, none against, 41 abstentions.

7. In France, for example, access to safe water and sanitation services remains an issue for 200 000 homeless people and 3 million urban residents with unhealthy housing conditions.

8. According to the definition provided in the European Directive on wastewater treatment, a sensitive area is a water body where further treatment than that prescribed in Article 4 of the Urban Waste Water Treatment Directive is necessary, either because it is eutrophic or at risk of becoming eutrophic in the near future, or because such treatment is necessary to fulfil other Council Directive(s).

9. Greenfield developments are defined as the creation of planned communities on land not yet converted to development (typically on green surroundings of cities). They contrast with brownfield developments, which occur on previously developed land.

10. Such a practice can be compatible with the "polluter pays" principle when compensations are attached to concrete objectives regarding the improvement of water quality and the sustainability of particular services.

11. http://stats.oecd.org/Index.aspx?Datasetcode=CITIES.

12. When the city population changes, the index measures the increase in the built-up area relative to a benchmark where the built-up area would have increased in line with population growth. The SI index is equal to zero when both population and built-up area are stable over time. It is bigger (lower) than zero when the growth of the built-up area is greater (smaller) than the growth of population, i.e. the city density has decreased (increased) (OECD Metropolitan database).

Annex 1.A1

Two city typologies relevant to water management

Applying the concept of urban metabolism,[1] Fernandez (2014) proposes a global typology of cities. The typology builds on the consumption of major categories of material and energy resources, including total energy, electricity, fossil fuels, industrial minerals and ores, construction minerals, biomass, water and total domestic material consumption. It comprises five types of cities:

- *Developing regions*: the most challenging cities in developing regions consume small quantities of materials and resources.
- *Significantly industrialising economies*: cities with access to abundant resources consume little electricity and score higher on other dimensions. Some cities in OECD countries, including Istanbul, Mexico and several Japanese cities (Nagoya, Osaka, Tokyo and Yokohama), belong to this category.
- *Transition economies*: cities featuring low consumption of industrial minerals and ores, and medium to medium-high consumption of all other categories. Cities in OECD countries in this category include Barcelona, Budapest, Lisbon, London, Madrid, Milan and Rome.
- *Mining and coal-fed economies:* mining and coal-fed economies have a specific pattern and are characterised by a high consumption of fossil fuels and lower consumption of biomass, construction minerals and domestic materials. Cities in OECD countries in this category include Athens, Berlin, Dublin, Prague, Santiago, Tel Aviv and Warsaw.
- *Low-density and high-affluence cities:* low-density and high-affluence cities consume high levels of material and energy sources; some face challenging climates. Cities in OECD countries in this category include Boston, Denver, Detroit, Melbourne, Phoenix, Sydney and Vancouver.

Water is not a discriminating factor in Fernandez's typology, suggesting that urban water management is affected by dimensions that are not specific to the water sector. These include endowment in energy and material resources, which affect technology choices for urban water management and incentives for energy-efficient water services.

Van der Steen (2011) proposes a typology of cities that is relevant to urban water management, based on features combining water issues with governance capacity, i.e. affordability; capacities in the water sector; rainfall patterns (tropical, moderate, or affected by climate change); scarcity of water resources; and potential for reuse of treated wastewater. Combining these distinctive features, van der Steen clusters cities into three broad categories:

- Type 1 – water management driven by basic service issues

- Type 2 – water management driven by water scarcity
- Type 3 – water management driven by climate change effects on rainfall patterns, flooding and water quality.

Note

1. Urban metabolism is defined as the study of the physical flows required to serve the urban economy.

References

Ahiablame, L.M., B. A. Engel and I. Chaubey (2012), "Effectiveness of Low Impact Development Practices: Literature Review and Suggestions for Future Research", *Water Air Soil Pollution,* Vol. 23, pp. 4253-4273.

Allan, T. (1997), "Virtual water: a long term solution for water short Middle Eastern economies, British Association Festival of Science – Water and Development Session", 1997 Leeds, United Kingdom, http://lwrg.org/water-food-and-trade.html.

Barraqué B. and L. Isnard (2013), "Les recherches sur la ville durable : l'exemple de la gestion durable des services d'eau", in *Urbanisme et services publics urbains : l'indispensable alliance*, ASTEE, Nanterre, France.

Bazoglu, N. (2011), "Measuring and coping with urban growth in developing countries", *Global Urbanization.*

Beal, S., S. Trow and B. Haywood Smith (2012), *Environment Agency, Ofwat, Defra Review of the calculation of sustainable economic level of leakage and its integration with water resource management planning. Contract 26777,* Strategic Management Consultants, Northumberland, United Kingdom.

Bhatia, R. and M. Falkenmark (1993), "Water resource policies and the urban poor: innovative approaches and policy imperatives", *Water and Sanitation Currents*, The World Bank, Washington, DC.

Bicheron P., et al. (2008), *GlobCover 2005 – Products description and validation report*, Version 2.1, available on the ESA IONIA website, http://ionia1.esrin.esa.int/.

Bolognesi, T. (2014), "Analyse structurelle des systèmes hydriques urbains en Europe: aspects organisationnels et défis patrimoniaux", *Revue d'économie industrielle*, Vol. 147, pp.51-86.

Booz Allen Hamilton (2007), "Lights! Water! Motion!", *Strategy+business magazine (Booz & Company),* spring 2007, Issue 46.

Brewer, M. (2014), *U.S. Drought Monitor: October 21, 2014* [online], National Drought Mitigation Center, Lincoln, Nebraska, http://droughtmonitor.unl.edu/ (accessed on 30/10/ 2014).

Clothier, B., et al. (2009), *Virtual Water: Emerging Issu*es, Royal Society of New Zealand, September 2009.

Davis, P. et al. (2013), "A selection framework for infrastructure condition monitoring technologies in water and wastewater networks", *Expert Systems with Applications,* Vol. 40, pp. 1947-1958.

European Commission (2012), *Water Scarcity & Droughts – 2012 Policy Review – Building blocks Non-Paper*, European Commission.

European Environment Agency (2013), *Assessment of cost recovery through pricing of water*, Technical report No 16/2013, Luxemburg, DOI:10.2800/93669.

Fernandez, J. E. (2014), "Urban Metabolism: City Typologies", presentation at 2014 MIT Europe Conference, Brussels.

FOEN (2012), *Mikroverunreiningungen: Spezialfinanzierung für ARA-Ausbau in der Vernehmlassung*, Federal Office for the Environment, Bern.

Frone, S. and D.F. Frone (2014), "Challenges in Analyzing Correlation between Water Infrastructure and Economic Development", *Procedia Economics and Finance,* Vol. 10, pp. 197-206.

Gaffney, O. and F. Pharand-Deschenes (2012), "Water in the Anthropocene. Welcome to the Anthropocene: A planet transformed by humanity", www.anthropocene.info.

Gerrity, D. and S. Snyder (2011), "The economic value of water in metropolitan areas of the United States", *Water Policy*, Vol. 13, pp. 443-458.

Grey, D. and C.W. Sadoff (2006), "Water for Growth and Development", in *Thematic Documents of the IV World Water Forum*, Comision Nacional del Agua, Mexico City.

Grey, D. and C.W. Sadoff (2007), "Sink or Swim? Water security for growth and development", *Water Policy,* Vol. 9, pp. 545-571.

Grimsey, D. and M.K. Lewis (2002), "Evaluating the risks of public private partnerships for infrastructure projects", *International Journal of Project Management,* Vol. 20, pp. 107-118.

Hallegatte, S. et al. (2013), "Future flood losses in major coastal cities", *Nature Climate Change,* Vol. 3, pp. 802-806.

Hunger, M. and P. Döll (2008), "Value of river discharge data for global-scale hydrological modelling", *Hydrology and Earth Systems Science*, Vol. 12(3), pp. 841–861.

IPCC (2007), *Climate Change 2007: Synthesis Report*, An Assessment of the Intergovernmental Panel on Climate Change.

Jenerette D., L. Larsen (2006), A Global Perspective on Changing Sustainable Urban Water Supplies, *Global and Planetary Change*, Vol. 50, pp. 202-211.

Kandulu, J. M., J. D. Connor and D.H. MacDonald (2014), "Ecosystem services in urban water investment", *Journal of Environmental Management,* Vol. 145, pp. 43-53.

Karaca, F. et al. (2014), "A comparative analysis framework for assessing the sustainability of a combined water and energy infrastructure", *Technological Forecasting and Social Change,* in press.

Kenway, S. (2014), "Lessons from the Australian Millennium drought, and emerging frameworks and management tools", communication at the Energy for Water Workshop organised by KAPSARK, Paris, 15 September.

Lloyd Owen, D. (2011), "Infrastructure Needs for the Water Sector", unpublished paper commissioned by the OECD in preparation of the OECD report *Water Security for Better Lives*, OECD Publishing, Paris.

London Assembly (2003), *London's water supply*, a report by the London Assembly's Public Services Committee.

Lorrain, D. and F. Poupeau (2014), "Ce que font les protagonistes de l'eau", *Actes de la Recherche en Sciences Sociales*, Vol. 203, pp.4-15.

Marsden Jacob Associates (2006), "Securing Australia's Urban Water Supplies: Opportunities and Impediments: A discussion paper prepared for the Department of the Prime Minister and Cabinet", www.environment.gov.au.

McDonald R.I. et al. (2014), "Water on an urban planet: Urbanization and the reach of urban water infrastructure", *Global Environmental Change*, Vol. 27, pp. 96-105.

McDonald, R.I., D. Shemie (2014), *Urban Water Blueprint: Mapping conservation solutions to the global water challenge*, The Nature Conservancy, Washington, DC, http://water.nature.org/waterblueprint/#/intro=true.

Ministry of Health, Labour and Welfare (MHLW) (2010), www.mhlw.go.jp/shingi/2010/02/dl/s0202-8g.pdf (in Japanese).

Mikovits, C., W. Rauch and M. Kleidorfer (2014), "Dynamics in Urban Development, Population Growth and their Influences on Urban Water Infrastructure", *Procedia Engineering,* Vol. 70, pp. 1147-1156.

National Research Council (2012), *Corps of Engineers Water Resources Infrastructure: Deterioration, Investment, or Divestment?*, The National Academies Press, Washington, DC.

OECD (forthcoming a), *A Framework for Urban Policies*, OECD Publishing, Paris.

OECD (forthcoming b), *Water Governance in Cities*, OECD Publishing, Paris.

OECD (forthcoming c), *The Governance of Water Regulators*, OECD Publishing, Paris.

OECD (forthcoming d), *Water Governance in Brazil*, OECD Publishing, Paris.

OECD (2014), *Water Governance in the Netherlands: Fit for the Future?*, OECD Publishing, Paris, http://dx.doi.org/10.1787/9789264102637-en.

OECD (2013a), *New Modes of Water Supply and Sanitation Management and Emerging Business Models*, OECD, Paris, www.oecd.org/officialdocuments/publicdisplaydocumentpdf/?cote=ENV/EPOC/WPBWE/RD%282013%297&docLanguage=En.

OECD (2013b), *Water Security for Better Lives*, OECD Publishing, Paris, http://dx.doi.org/10.1787/9789264202405-en.

OECD (2013c), *Water and Climate Change Adaptation. Policies to Navigate Uncharted Waters,* OECD Publishing, Paris, http://dx.doi.org/10.1787/9789264200449-en.

Kim, S.-J., A. Schumann and R. Ahrend (forthcoming), *What Governance for Metropolitan Areas?*, OECD Regional Development Working Papers, OECD Publishing, Paris, http://dx.doi.org/10.1787/9789264231092-en.

OECD (2012a), *OECD Environmental Outlook to 2050: The Consequences of Inaction*, OECD Publishing, Paris, http://dx.doi.org/10.1787/9789264122246-en.

OECD (2012b), *Redefining "Urban": A new way to measure metropolitan areas*, OECD Publishing, Paris, http://dx.doi.org/10.1787/9789264174108-en.

OECD (2012c), *Compact City Policies. A Comparative Assessment,* OECD Publishing, Paris, http://dx.doi.org/10.1787/9789264167865-en.

OECD (2010), *Cities and Climate Change*, OECD Publishing, Paris, http://dx.doi.org/10.1787/9789264091375-en.

OECD (2006), *Infrastructure to 2030. Telecom, Land Transport, Water and Electricity*, OECD Publishing, Paris, http://dx.doi.org/10.1787/9789264023994-en.

Ofwat (2012), *Distribution input and supply pipe leakage for England and Wales: Dataset for 1992/93 to 2010/11* [Online], http://data.gov.uk/dataset/distribution-input-and-supply-pipe-leakage (accessed on 31/10 2014).

ONEMA (2012), *Observatoire des services publics d'eau et d'assainissement. Panorama des services et de leur performance*, Paris.

PBL (Netherlands Environmental Assessment Agency, 2014), *Towards of World of Cities in 2050. An Outlook on Water-Related Challenges*, background report to the UN-Habitat Global Report.

Postel, S. (1997), *Last Oasis: Facing Water Scarcity,* Worldwatch Institute, New York.

Poston, B. (2015), "Aging water mains a $1 billion headache for DWP", *Los Angeles Times*, 16 February, 2015.

Ray, C. and R. Jain (2014), "Chapter 7 – Water Infrastructure Development for Resilience", *in* Jain, C. R. (ed.), *Low Cost Emergency Water Purification Technologies,* Butterworth-Heinemann, Oxford.

Richter, B.D. (2013), "Tapped out: how can cities secure their water future?", *Water Policy*, Vol. 15, pp. 335-363.

Rodriguez, C. et al. (2009), Indirect Potable Reuse: A Sustainable Water Supply Alternative, *International Journal of Environmental Research and Public Health,* 6(3), pp. 1174-1209.

Rouse, M. (2006), *Sustainable Cost Recovery*, IWA Institutional Governance and Regulation Specialist Group and IWA Statistics and Economics Specialist Group, IWA Reference Paper, August 2006.

Sanders, K. and M.E. Webber (2013), "The Energy-Water Nexus: Managing Water in an Energy Constrained World", *Earth Magazine*, July 2013.

Sitzenfrei, R. and W. Rauch (2014), "Investigating Transitions of Centralized Water Infrastructure to Decentralized Solutions – An Integrated Approach", *Procedia Engineering,* Vol. 70, pp. 1549-1557.

Steen P. (van der) (2011), *Application of Sustainability Indicators within the Framework of Strategic Planning for Integrated Urban Water Management*, www.switchurbanwater.eu/outputs/pdfs/W1-1_CALE_MAN_D1.1.7_Indicators_Manual.pdf.

Tortajada, C. (2006), "Water management in Singapore", *International Journal of Water Resources Development,* Routledge, Vol. 22, pp. 227-240.

USEPA (2013), *The Importance of Water to the U.S. Economy. Synthesis report*, US Environmental Protection Agency, Washington, DC.

USEPA (2013b), *Drinking Water Infrastructure Needs Survey and Assessment. Fifth Report to Congress*, US Environmental Protection Agency, Washington, DC.

Water UK (2011), *Sustainability Indicators 2010-11*, London.

Chapter 2

Financing urban water management

This chapter examines the challenges related to financing urban water management. Several factors drive financing needs up when traditional sources of finance are constrained. For example, declining water use per capita in several city centres can positively affects water conservation, but negatively affects revenues from water tariffs, as well as the service provider's financial capacity to operate and maintain the existing infrastructure.

The chapter reviews options (e.g. green infrastructures) cities can explore to minimise costs, make the best use of tariffs and taxes, diversify revenue streams and access private financing. It considers the role of financiers, property developers and other private actors.

Key messages

Keeping up with current levels of water security in cities in OECD countries will require sustainable financing to upgrade and renew existing infrastructures, and eventually build new assets. Yet prevailing sources of financing for urban water management are, and will likely remain, severely constrained: – declining per capita water consumption in several city centres lowers water-tariff revenues, fiscal consolidation reduces the public budgets available for infrastructure investment, and international solidarity mechanisms are unlikely to benefit cities in OECD countries.

Robust financing strategies for urban water management in OECD countries combine four elements. First, they minimise operating costs and investment needs through targeted maintenance, efficiency gains (e.g. from amalgamating water services at the right scale), or using low-cost water resources. Each option requires specific institutional arrangements, e.g. amalgamation requires organising deliberative and technical bodies at the appropriate scale; water reuse requires specific regulations on the quality of reclaimed water.

Second, robust financing strategies explore tariff structures that contribute to water resource management (particularly water conservation) and the financial sustainability of water services. Three interrelated questions deserve renewed attention:

- How to compensate for declining water demand? The chapter reviews *(i)* innovative tariff structures that partially decouple revenues for service providers from the volumes of water sold; and *(ii)* performance-based contracts that reward attaining specific objectives, e.g. reducing non-revenue water or conserving water or energy.
- How to address affordability issues? The distributional effects of sophisticated tariffs require thorough analysis to avoid financial losses and inequitable consequences. The chapter reviews several responses that combine targeted and time-limited support for those who need it and incentives to use water-saving appliances.
- Who should pay the water bill? People and actors who benefit from urban water management or generate additional costs should be identified and harnessed to foot the bill. Fiscal instruments can redress externalities associated with land development or impervious surfaces.

Third, revenue streams for water management can be more diverse when cities consider new fiscal instruments (e.g. taxes on land or impervious surfaces), or when utilities develop new services. Finally, cities can tap into new sources of capital. The private sector, including financiers, property developers and small entrepreneurs, is gaining experience in financing discrete facilities (desalination or wastewater treatment plants, distributed infrastructures) at different scales. Public utilities, for their part, recycle some of the capital tied up in water infrastructures to generate cash for use in new projects. National and local governments need to explore innovative ways to jumpstart and leverage private investment where required.

Innovation around these issues abounds in cities, but some institutional barriers – including the preference for incumbent technologies and cheap water (which fails to recognise negative externalities, such as poor maintenance or degraded performance of water services), as well as political interference (e.g. service micromanagement, sub-optimal and unstable pricing policy, and biased or incomplete definition of performance for water services) – need to be overcome in order to diffuse it more systematically. The three subsequent chapters address some of these barriers.

Introduction

This chapter analyses how financing needs for urban water management in OECD countries can be met. It first documents trends in traditional sources of finance for urban water services, including tariffs, taxes and transfers (the 3Ts) and repayable finance. It then explores the strategies cities in OECD countries can use to cover their financing needs; these combine options to minimise costs (e.g. through targeted maintenance or amalgamated services), innovative tariff structures and additional financing sources (e.g. innovative business models for water utilities and taxes that harness beneficiaries of enhanced water security). It covers both investment in, and operation and maintenance of, urban water infrastructure and services.

The chapter highlights the connections with innovative technologies to manage water and the potential benefits of co-ordination with the rural environment. It further identifies requisites for facilitating the exploration of innovative and financially sustainable options for urban water management, including regulatory policy, institutional arrangements and stakeholder engagement.

The Republic of Korea illustrates some of the main points developed in the chapter, with supporting evidence (see Annex 2.A1)which confirms that: *(i)* as water infrastructure coverage develops, operation and maintenance represent a significant and rising share of total costs for urban water management, raising raises specific financing issues; *(ii)* heavy reliance on public finance is not sustainable, particularly in the context of the economic and financial crisis, where public financing is scarce and competition fierce; *(iii)* revenues from water bills are lowered by stable or declining water consumption; and *(iv)* new sources of finance can be explored, drawing on people and actors who benefit from investment in wastewater collection and treatment but do not pay the cost.

Financing constraints in cities in OECD countries

Conventional economic theory argues that revenues from user charges should cover all the costs related to the service: operations and maintenance (O&M) costs, capital costs and the opportunity costs of using the resource (which can be significant when there is competition to access water sources). The OECD has a distinctive approach acknowledging three sources of funds (the 3Ts) that ultimately finance water services and infrastructures: tariffs (revenues from user charges), taxes (subsidies and transfers from public budgets) and transfers from the international community. Other sources of financing – repayable finance, including loans, bonds and equity – can cover upfront investment, but will be repaid through a combination of the 3Ts.

This section reviews general trends in financing for water services and infrastructures. It focuses on the 3Ts and on repayable finance (essentially bank loans and private equity). General barriers to domestic and international investment (e.g. regulatory, currency and corruption risks) apply in the water sector, but are not covered in this report.

The section argues that cities in OECD countries face growing limitations to accessing the revenues and capital they need to invest in water-related infrastructures. These trends are not limited to large cities. In New Zealand, small towns experience equal hardship due to population declines and ageing infrastructure, leaving a reduced rate base (local taxation) affecting local governments' ability to fund, operate and maintain infrastructure.

Revenues from water tariffs

The OECD (2010a) argues that water tariffs serve four potentially conflicting objectives: financial sustainability, economic efficiency, environmental sustainability and social equity (Figure 2.1). Trade-offs between the four objectives and the capacity to address them evolve over time: income improvements may enable a low-income community to pay the prices needed to obtain previously unaffordable services; technological improvements may reduce costs; more effective institutions may emerge; social learning processes may enable the community to accept previously unacceptable solutions (e.g. pricing). It follows that pricing strategies would benefit from recurrent assessment and revision.

Figure 2.1. **Policy objectives and trade-offs affecting price structures for water services**

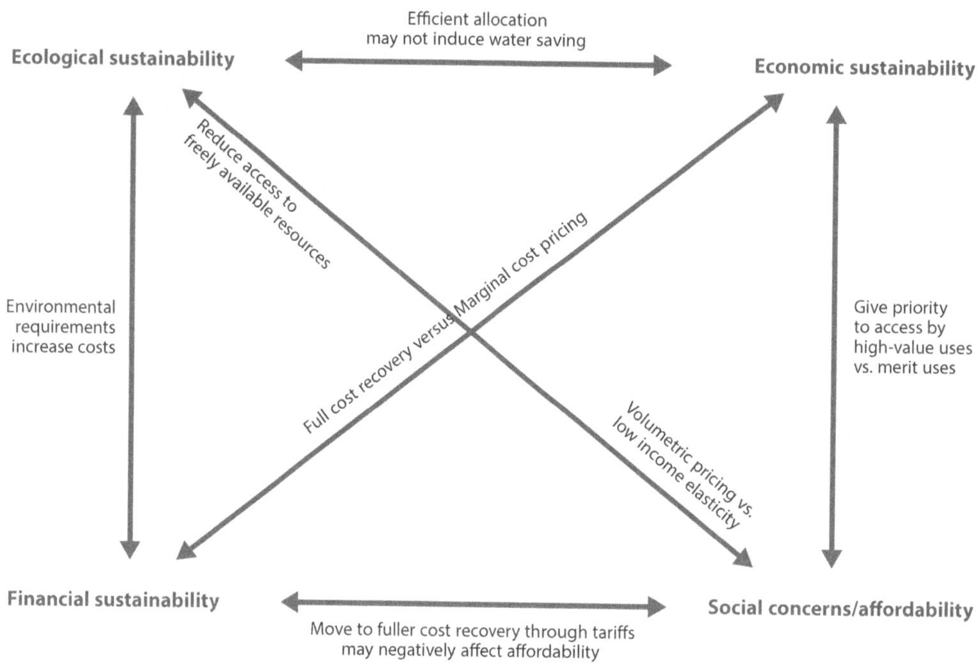

Source: OECD (2010a), *Pricing Water Resources and Water and Sanitation Services*, OECD Publishing, Paris, http://dx.doi.org/10.1787/9789264083608-en; based on Massarutto, A. (2007b), "Abstraction Charges: How Can the Theory Guide Us?", presentation at the OECD Expert Meeting "Sustainable Financing for Affordable Water Services: From Theory to Practice", November 2007.

Revenues from user charges are based on the volume of water used and/or wastewater collected (or a proxy thereof) and the rates per unit of water used or treated. They are affected by urban water use per capita and by inefficiencies in the tariff-setting process, especially when a disproportionate portion of costs incurred by service providers is fixed. Note that the share of variable costs is higher when utilities procure bulk water from wholesalers.

Per capita water consumption is declining in cities in OECD countries due to a combination of increased system efficiency, deindustrialisation and lower levels of domestic use. Per capita consumption decreased by 25% in Paris over 15 years, by 16.5% in Berlin between 1995 and 2005 (Poquet and Maresca, 2006) and by 31% in Nantes (France) between 2003 and 2008 (Figure 2.2). The OECD survey on water governance for future cities (OECD, forthcoming a) confirms this trend. Water consumption has declined generally over the last ten years, dropping sharply in cities like Acapulco, Budapest, Culiacan

(Mexico), New York and Phoenix; only Mexico City and Hong Kong have registered a slight increase in water consumption compared to their 2000 levels, indicating that reduced water consumption per capita becomes the "new normal" in cities in OECD countries. This decline contrasts sometimes with trends in peri-urban areas, where urbanisation is driving water consumption.

Figure 2.2. **Water consumption in selected cities**

2000-12; square metre (m³)/inhabitant/year

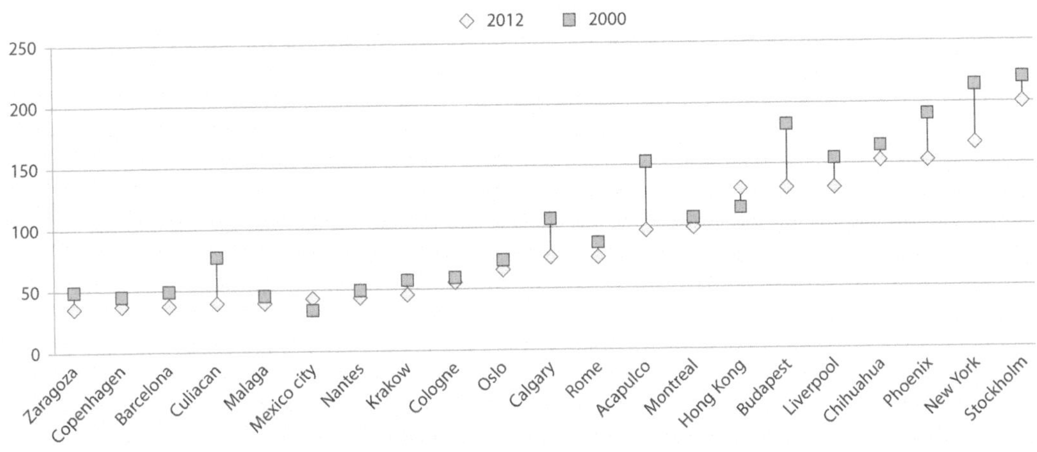

Note: the sample includes cities that provided data for 2012 and 2000.

Source: OECD (forthcoming a), *Water Governance in Cities*, OECD Publishing, Paris.

A combination of factors are driving this trend. In France, domestic water consumption per capita has been declining by 1% a year since 2000 (ONEMA, 2012), driven by the relatively lower number of industrial activities in urban settings; water demand management in collective buildings to minimise costs; innovation and more water-efficient new appliances; higher water prices; and households' attitudes to the environment and wastage.

- Industrial uses are declining, driven by efficiency gains and the shift from industry to services. Similarly, hospitals and schools tend to reduce their water consumption.
- Households have also reduced their consumption (see Figure 2.3 on Japan). While per capita water consumption initially dropped due to reduced outdoor use, indoor water use (chiefly toilet-flushing and washing machines in the United States; see de Oreo, 2010) has also decreased.
- A growing number of operators are seeing an additional decrease in consumption because households are finding alternative water supply solutions – e.g. rainwater harvesting and private well-drilling (OECD, 2013a) – driven by environmental motives (reducing the energy footprint) or financial concerns (reducing the water bill). These households usually remain connected to the main infrastructure, and in some cases discharge water they have not purchased into the sewer, free-riding on the sewer service.
- Authorities contribute to this trend when they encourage users to conserve water. In California, the 2009 Water Conservation Act requires the state to achieve a 20% reduction in urban per capita water use by the end of 2020.

Figure 2.3. **Domestic water consumption in Japan**

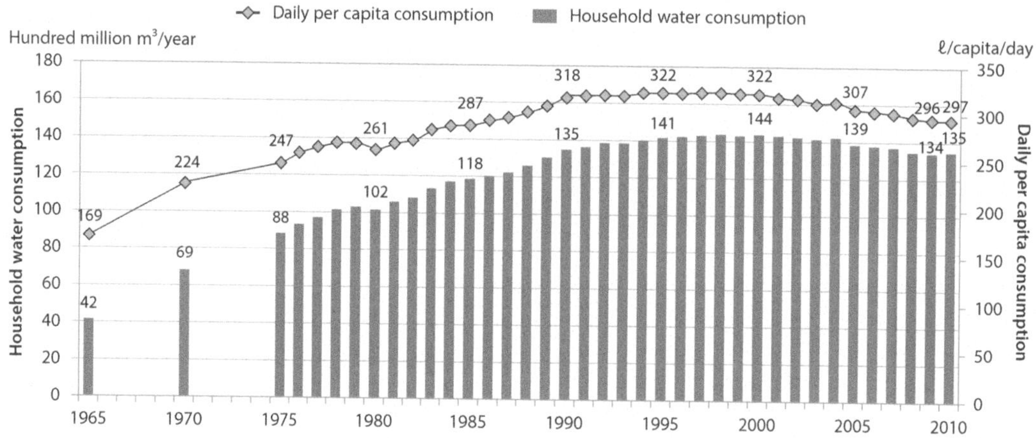

Source: Ministry of Land, Infrastructure, Transport and Tourism (2013), "Water Resources in Japan" (in Japanese).

Whether per capita water consumption in cities in OECD countries has reached its lower limit, or will decline even further, is unclear. Projections on future water demand in cities in OECD countries do not always anticipate the shifts in actual water consumption. The Oxford Water Futures Programme studied 21 projections for future water-supply needs in England and Wales made between 1949 and 2009, plotting them against the actual amount put into supply over the years. In the mid-1970s, the utilities were supplying some 13 billion litres a day. Two projections in the 1970s foresaw a requirement of 28 billion litres a day in 10 to 20 years' time, two other projections anticipated a requirement of 24 billion litres and one projection foresaw only 20 billion litres. In fact, supply peaked at 17 billion litres over 1949-2009 and is back today at 14 billion litres (Lloyd Owen, 2011).

Declining water consumption has mixed consequences (Conseil national de l'eau, 2013). On the one hand, it relieves pressure on the resource and allows coping with future uncertainties without additional investment. On the other hand, it raises technical, social and political issues:

- It slows the water flowing in pipes and generates quality risks.
- It affects (public or private) utilities' revenues when they are based on the volumes of water sold or treated. As water services are characterised by a high share (up to 80%) of fixed costs, a disproportionate price increase is required to compensate for lost revenues from enhanced water conservation.[1] San Francisco (California) has experienced this spiral of drought-triggered conservation measures, driving water consumption down; in response, rates for water supply increased by 5.9% in 2014.

The mixed consequences of declining water consumption explain why the city of Hamburg ceased to promote water efficiency and conversation as technical and financial concerns about further decreases in water consumption overrode the potential benefits of water-saving. It follows that conservation measures should be experimented first in areas where water scarcity or growing water needs have brought existing infrastructure to full capacity, and where additional water supply would be very costly.

One option to compensate for declining water consumption is to raise tariffs. However, there are limits to water-related price increases. First, while the vast majority of city dwellers in OECD countries could afford larger water bills, rising water tariffs have led to water-poor households. The OECD survey on water pricing (OECD, 2010a) establishes that water bills represent 3%[2] or more of households' budgets for the lowest decile of the population in several OECD countries (Figure 2.4). In England and Wales, the amount of water payments in arrears increased sharply after the water supply was privatised and disconnecting non-payers became illegal. While most bad payers are not paying on principle rather than from a lack of means, some cannot afford to pay. The last section of the chapter discusses tariff structures and other measures to compensate for declining water demand and conservation policies while addressing affordability issues.

Figure 2.4. **Water supply and sanitation bills as a share of disposable income of the lowest decile of the population**

Selected OECD countries. Average income of the lowest decile of the population, 2008

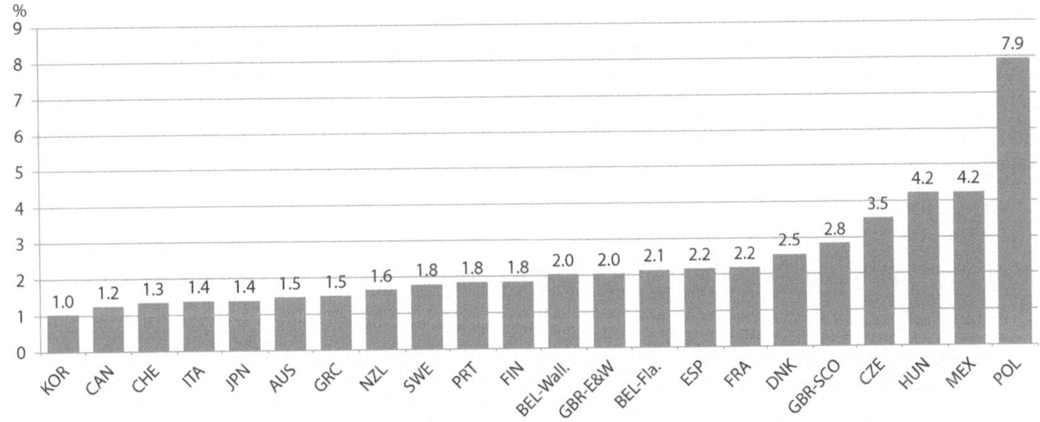

Source: OECD (2010a), *Pricing Water Resources and Water and Sanitation Services*, OECD Publishing, Paris, http://dx.doi.org/10.1787/9789264083608-en.

Second, higher tariffs associated with declining water consumption can confuse customers, who may perceive them as a punishment for virtuous behaviour. This can explain why political entities tend to avoid raising rates to recover lost revenues. In France, price increases do not fully reflect the decline in the volume of water sold, thanks to productivity gains and the decreased financial burden of utilities after repaying former loans (Conseil national de l'eau, 2013). Whether the reduced revenues affect service quality, infrastructure maintenance and investment flows, however, is unclear.

The trends analysed above challenge the business model of urban water management. It is widely accepted that the costs of the service should be covered by the water bills[3] (OECD, 2010a). However, the sustainability of this model is questionable as costs increase and volumes of water sold diminish. Moreover, the model may not comply with core principles of financing water services (OECD, 2012). For instance, some beneficiaries of improved water services may not contribute to the costs, or the water bill may cover expenditures generated by other users – as when water bills cover the costs of rainwater management or the treatment of diffuse pollution in groundwater. In this context, cities are encouraged to revisit what services the water bill should cover – subsequent sections of the chapter explore this in more detail.

Taxes (transfers from public budgets)

Taxes are the other main source of financing for urban water services, essentially to balance costs and revenues at the national and/or sub-national levels. Access to public budgets is severely constrained, particularly since the financial crisis triggered fiscal consolidation policies. Kaminker et al. (forthcoming) note that the economic and financial crisis has constrained government budgets in many OECD countries, exerting downward pressure on public sources of investment financing for infrastructure. Furthermore, fiscal consolidation efforts to reduce the share of government debt in the gross domestic product have been accompanied in some countries by pressure to reduce support for urban infrastructure. Ireland's recent shift from a model where taxpayers paid for water services to one where users (industry and households) pay for it was part of the International Monetary Fund-European Union bailout programme to save public finance and meet important investment needs.

Two avenues for financing urban water services are being considered. One is to optimise the leverage effect of public finance. In the United States, federal support for water infrastructure has gradually changed. In the 1970s, the federal government provided large grants to finance sewage treatment plants. Starting in 1987, the Clean Water State Revolving Fund rotating loan programme superseded the grants, followed by the Drinking Water State Revolving Fund in 1996. Between their inception and the end of 2011, the funds have helped finance USD 111 billion (US dollars) worth of infrastructure projects (Walton, 2012). Since 2011, local governments cover nearly the entire capital cost of water infrastructures.

This strategy faces limitations in the current context. Donnelly and Christian-Smith (2013) note that while municipal bonds were historically a relatively inexpensive way of financing new infrastructure, credit rating agencies today are increasingly downgrading municipal water systems. The authors quote a report by Ceres that blames these downgrades principally on the fact that water tariff increases have not kept pace with spending on system maintenance or debt service coverage.

Another popular financing avenue is to arrange public support that does not translate into public debt (see Box 2.1 for an illustration in the United States).

Box 2.1. **Innovative public finance support in the United States**

The United States Congress is considering a water-only loan programme (currently a draft, accessed on October 2014) whereby the Water Infrastructure Finance Innovations Authority would provide low-interest loans to be repaid at long-term US Treasury rates. This programme could grant loans for large infrastructure projects – which can hardly be financed by state revolving funds – thereby lowering the cost of access to capital for water utilities.

This programme would have little effect on US public debt, as only the money considered at risk of default is counted as federal expenditure, and default in water projects is very low.

Source: Circle of Blue (2012), "America's Water Infrastructure shows its Age", www.circleofblue.org/waternews/2012/world/americas-water-infrastructure-shows-its-age-the-national-debate-about-how-to-pay-for-repairs/ (accessed February 2014).

Transfers from the international community

Transfers from the international community play a marginal role in financing urban water management in OECD countries. One notable exception is the EU Structural and Cohesion Funds, designed to support upgrading urban infrastructures in new Member States or those struggling to meet EU regulations. The funds provide a total of EUR 8 billion to help reduce leakage rates, connect to the water supply, generate additional supply and improve infrastructure, and an additional EUR 7 billion for disaster prevention. In addition, the Solidarity Fund allocates around EUR 1 billion a year for disaster relief in Member States.

Other international mechanisms may emerge as potential sources of financing for urban water management. Carbon finance already finances investment in wastewater treatment in non-Annex I countries[4] through Clean Development Mechanisms. Water is gaining traction in national and international climate policies: it is playing a critical role in climate change adaptation as it becomes increasingly clear that climate change will result in significant changes in water availability and variability, as well as water-related weather events such as droughts and floods. However, the precise manner in which climate financing can be mobilised to enhance urban water management in OECD countries is unclear.

Recent OECD work on the role of water in climate change adaptation established that most OECD countries have yet to address adequately the financing required to adapt water systems. Their national adaptation strategies and plans only briefly – if at all – address financing issues (OECD, 2013b). A few OECD countries have relied on international funding mechanisms to support adaptation activities. Chile has received support to develop climate change-related activities from the Global Environment Facility, its implementing agencies and bilateral development co-operation partners. In 2010, the World Bank approved a USD 450 million loan to Mexico to support government policies fostering preparedness to confront the growing impacts of climate change through programmes by the National Water Commission.

Repayable finance

Urban water management is struggling to attract private capital. As noted in OECD (2006), the profile of urban water investment projects typically involves a high initial capital outlay, followed by a very long payback period from long-lived assets. As a result, the risk profile of water investments compares unfavourably with many other projects. The information asymmetry between governments and water utilities, and the political sensitivity of water pricing, leave the sector vulnerable to ad hoc politics (unwillingness to charge) and social criticism. Consequently, the sector often suffers from a high level of political interference and confusion over its social, environmental and commercial aims.

Recent trends in private financing make access to public funding for urban water management even more challenging. In the wake of the economic and financial crisis, sources of investment financing in the corporate (e.g. utilities) and financial sectors face significant constraints, and may continue to diminish in the coming years.

Utilities have insufficient capacity to expand their investment in water infrastructures. Both public and private utilities' balance sheets are limited by the negative impacts that any debt increase would have on their credit rating and cost of capital. Industry leaders (e.g. Veolia and Suez Environnement) have radically revised their approach to investing in new contracts, as evidenced by their renewed preference for management contracts (instead of concessions), where they do not have to invest their own capital to develop infrastructure.

The capacity of cities and utilities to access bank loans is unevenly spread across countries. It has decreased in many OECD countries as a consequence of the 2008 financial crisis, which prompted banks to reduce investments across illiquid asset classes and shorten the term of loans (Kaminker et al., 2015). Local authorities are particularly affected, since governments set limits on how much a city can borrow. Access to credit becomes a distinctive feature of private and public operators. Under certain circumstances (e.g. where public authorities are heavily indebted), obtaining credit may be easier for private operators, making the case for management contracts or public-private partnerships. Under other circumstances, public authorities may have better access to credit than private operators.

Given the context, private investors will be even more selective when considering urban water-management projects. As with any other infrastructure projects (see Kaminker et al. [forthcoming] for a more detailed discussion pertaining to the renewable energy sector), urban water projects will only attract private capital if they can provide sufficient collateral, the probability of success and predictable future cash flows to guarantee investor income. The trends in revenues from water tariffs (analysed above) indicate this can be challenging; however, recent developments (see below) confirm that watër projects do attract private funds, including from purely financial investors.

Options to minimise costs

Chapter 1 showed that cities in OECD countries have displayed a marked preference for grey infrastructures, often managed at municipal level, to counteract different water risks. Cities have built reservoirs and dykes to protect against risks of droughts and floods; they have laid extensive piped networks to supply water, and collect and treat rain and wastewater.

Alternative modes of urban water management exist that require less capital investment and/or are less costly to operate and maintain. This section envisages options to minimise costs for urban water management. Some are technical and pertain to enhanced infrastructure management, alternative and cheap sources of water, and alternative technologies. Others are non-technical and relate to the organisation (and particularly the geographic scale) of urban water management. A number of these options rely on co-operation between cities and their rural environment, and entail adjusting the organisation and governance of urban water management. Subsequent chapters revisit these linkages.

Tapping efficiency gains in asset management

Increased efficiency in O&M of existing assets can be a cost-efficient way of improving water security and services. Urban utilities in developed countries increasingly rely on computer tools, inspection robots and geographical information systems (GIS) to gain a precise knowledge of the state and performance of their assets, particularly those buried underground (Box 2.2). This knowledge allows them to better phase their maintenance and renewal investments to improve system reliability (especially by repairing damaged pipes). Innovative tools help enlarge the scale and scope of infrastructure monitoring, and extend the time horizon for asset management.

A multidimensional definition of performance for water services – from basic physical condition to service quality, business risk and sustainability – can underlie methods to optimise the management of existing infrastructure. Tools to assess the state of infrastructure can bolster decisions to design financing strategies that factor in the impact of costs on water prices and the consequences of environmental, social and economic disruptions.

Box 2.2. **Advanced asset management: Illustrations from selected OECD countries**

In New Zealand, the city of Auckland used GIS to overlay actions and investments with a direct or indirect effect on freshwater quality, including those targeting:

1. stormwater asset maintenance, renewal and development
2. cycleway and road construction
3. network infrastructure development (e.g. broadband rollout).

The case study highlights the potential benefits of advanced asset management. It also suggests additional benefits deriving from co-ordination across such sectors as road construction.

In the United States, the Massachusetts Water Resources Authority developed a predictive maintenance strategy based on condition monitoring, and the probability and consequences of failure of each component. The programme increased equipment availability to 99%; it achieved cost savings by eliminating unneeded and low-value preventive maintenance work, and shifting the freed-up resources to predictive tasks and actual maintenance work. Predictive and probability-based maintenance illustrates a shift from zero-risk asset management (which translates into high degrees of infrastructure redundancy) to more-thorough risk analysis allowing more strategic and cost-effective asset management.

In the United Kingdom, an advanced pressure management system combining software, sensors and controllers is used to detect leakages early on, reducing water loss by 1.5 million litres a day.

Sources: case study developed by Andrew Schollum and Grant Barnes, Auckland Council, for this OECD project; Massachusetts Water Resources Authority; Water Environment Research Foundation; World Economic Forum (WEF) (2014), *Strategic Infrastructure. Steps to Operate and Maintain Infrastructure Efficiently and Effectively*, WEF, Geneva; i2O (2012).

When the operator and authority have a clear vision of asset and renewal needs, as well as an improved forecasts of water demands, they can rigorously plan operation, maintenance and investment. Moreover, they can sign precise and secure contracts that reduce information asymmetries and rent-seeking behaviour by either party (OECD, 2010b). Finally, rigorous asset management entails precise depreciation, which in turn leads to improved self-financing capacities, reduced debt and access to cheaper loans (since the utility is more creditworthy).

A private operator who is responsible for asset management can contribute by sharing knowledge and capacity. Clearly defined roles and responsabilities, as well as performance targets and contractual arrangements promoting efficiency, foster private-sector participation.

Exploiting low-cost water sources

Marsden Jacob Associates (2006) analyses the costs of major supply and demand options available to Australian cities (Figure 2.5). While the study dates and costs need to be updated, the main message remains: all things being equal, contextual features determine the cost advantage of any option.

- Most options (e.g. catchment thinning or purchasing irrigation water) have very low costs in favourable locations and situations.
- Many options (e.g. rainwater tanks or long distance pipelines) have very high costs (over USD 3.00 per kilolitre) in unfavourable locations and situations.
- The costs of pipelines and pumping have an overriding influence when water needs to be transported over long distances.

The driving force of transport costs has several consequences on infrastructure design and management. First, it explains why urban services face diseconomies of scale at the far end of existing infrastructures (Barraqué, 2003). Second, it suggests that wastewater reclamation is more cost-effective when treatment facilities are located close to potential (industrial, agricultural or municipal) users. On the one hand, major new water reuse initiatives are frequently comparably priced – or more expensive than – desalination due to the long transportation distances. On the other hand, decentralised systems are more advantageous when wastewater is treated and reused close to its collection point, thereby saving on transport costs, infrastructure and energy.

Figure 2.5. **Direct costs of alternative water supply options**

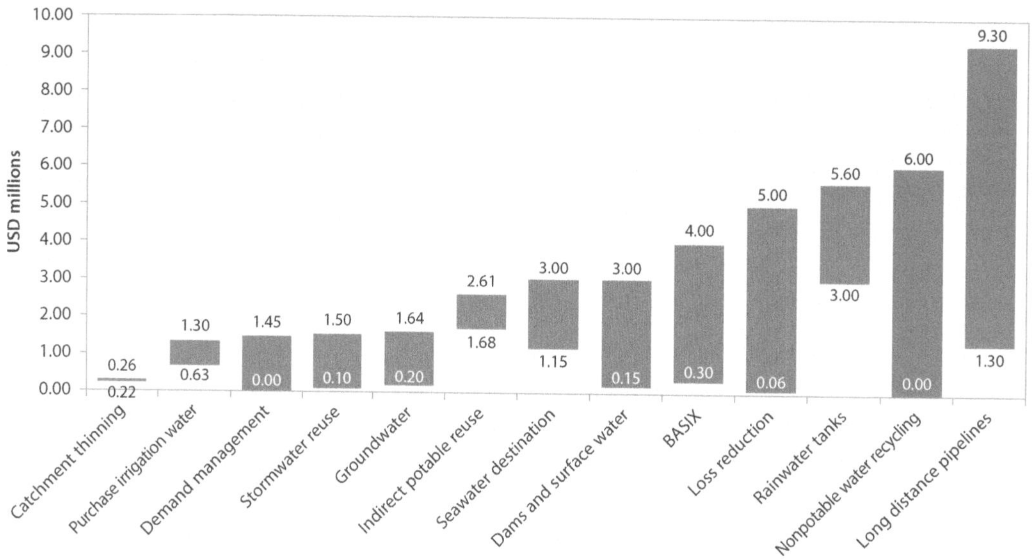

Source: Marsden Jacob Associates (2006), "Securing Australia's Urban Water Supplies: Opportunities and Impediments. A discussion paper prepared for the Department of the Prime Minister and Cabinet", www.environment.gov.au.

As Chapter 3 will show, innovative ways to manage water can reduce the costs of urban water management. Green infrastructure in particular can prevent or postpone the cost of building or extending grey infrastructures. For instance:

- Building new dams and rainwater harvesting facilities would cost USD 0.04-0.06 per cubic metre (USD/m^3), whereas rehabilitating existing infrastructure would only cost about USD 0.02/m^3; demand-side measures would be far cheaper (2030 Water Resources Group, 2009).
- In Philadelphia, the proposed eco-friendly "sponge-like" water system involving new forms of drainage (green roofs, wetlands, repaving with porous materials) would cost USD 2 billion, less than half as much as a conventional upgrade of the current pipe and basin system (WEF, 2014); achieving a similar level of service through an additional wastewater treatment plant would be 4 or 5 times more expensive at USD 8-10 billion (Walton, 2012).
- In Australia, a pilot project funded by Queensland Urban Utilities in partnership with SEQ Catchments to repair 500 metres of eroded riparian corridors near the Beaudesert

Sewage Treatment Plant in the Logan River catchment can achieve the same level of environmental performance as upgrading the treatment plant at a lower cost.

- In California, the East Bay Municipal Utilities District uses methane from waste to power generators; it is one of the first US wastewater treatment facilities to become a net-energy producer (US Environmental Protection Agency [USEPA], 2013a).

The costs of preventing the degradation of water quality should receive similar attention. Chapter 4 argues that catchment protection from diffuse pollution can be a cost-effective way to manage water quality.

Amalgamating urban water management

Non-technical options exist to minimise costs. Many OECD countries have aggregated (or are considering aggregating) small utilities to generate economies of scale and make the best use of large infrastructures. Heavy investment costs and the phasing out of government subsidies have prompted local utilities to concentrate part or all of the tasks related to the provision and delivery of WSS services at upper levels of government (OECD, 2013a; see also Chapter 1 for selected illustrations).

In New Zealand, the amalgamation of several councils gave the Auckland Council the necessary scale to tackle issues that were previously beyond the capacity of individual councils. Since amalgamation, the Council has been able to accelerate the modernisation of the region's antiquated wastewater treatment systems, substantially upgrade its two key wastewater treatment plants and progress the NZL 950 million (New Zealand dollar) "central interceptor" project that will reduce overflows from the combined waste and stormwater system of the Auckland isthmus. In Korea, cities in the Gyeongnam province achieved cost efficiency by amalgamating urban water services (Box 2.3).

Because the size and cost of sewer pipes are usually higher than those of the water supply infrastructure, sewerage units are less concentrated (OECD, 2013a). However, amalgamation may occur in wastewater treatment, where economies of scale exist for some complex treatment processes.

Amalgamation eventually results in combining different services at different scales. France's Ile-de-France region has a three-tier management system: street sewers are municipal, interceptors and storm sewers are run by the counties (four *départements*) and sewage treatment is operated by a joint-county (almost regional-level) board.

Several countries have separated water or treated wastewater production and the delivery of the service to customers:

- In Boston, a metropolitan authority consolidates water production and sewage treatment, leaving member municipalities in charge of system management.
- In Portugal, the government created a national water company in 1994. Municipalities in the same area were offered the opportunity to manage treatment plants jointly, while communes kept responsibility for operating water and sewer mains.
- In Australia, the 1994 reform planned by the Council of Australian Governments mandated the unbundling of former urban water monopolies, with bulk water production and sewage treatment organised at the regional level (by one public company) and retail water services at a more local level (by several water distribution companies). This choice paved the way for alternative water supply technologies (e.g. recycling and desalination).

Box 2.3. **Amalgamated water services – Gyeongnam Province, Korea**

Many Korean municipalities are having a hard time managing their own waterworks: the lack of revenue from low water tariffs leads to financial constraints on renewing existing water infrastructures. Ageing water infrastructures, particularly ageing water mains, are the predominant cause of water leakage, driving production costs and water tariffs up.

To solve these issues, the central government supports and encourages municipalities to amalgamate water supply services and assign amalgamated services to specialised water agencies.

Four local governments in southwest Gyeongnam Province amalgamated their water supply systems and assigned their operation to K-water. Each local government retains ownership of its water supply system and remains responsible for providing the service and setting its tariffs, as well as for planning and extending water mains in order to increase access to tap water. The tasks devolved to K-water include water abstraction and treatment, distributing treated water to customers, and notifying and collecting water tariffs.

K-water has installed an integrated remote-control centre to monitor and control each municipality's water sources, treatment plants and reservoirs. Most facilities, except those located far from city centres, have no staff. Operators of the integrated remote-control centre monitor water pressure and manage facilities 24 hours a day, 7 days a week. They are available at all times to respond immediately to calls from a facility. If the systems are out of order, engineers working for a local service centre are expected to be able to reach the facility within 30 minutes through a network of emergency contacts.

In order to enhance operational efficiency, K-water covers the upfront capital costs of renewing and upgrading ageing infrastructures. It charges each local government on a monthly basis for the operating expenses, including investment recovery. The contract specifies the amount to be paid by the municipalities, providing them with the ability to plan expenditures in advance.

The project has received positive reviews from the central government and municipalities involved. It is expected to cut costs by KRW 24 billion (Korean won) (EUR 19 million) over the contract duration (between 20 and 30 years), compared with business as usual. The volume of water accounted for has increased between 17.1% and 41.3% in the new system.

How amalgamation affects consignment charges

Name of municipality	Consignment charge[a] under individual consignment	Consignment charge under integrated consignment	Difference in consignment charge	Cost-cutting effect (%)	Note
Sacheon	749 KRW/m^3	727 KRW/m^3	22 KRW/m^3	2.9	30-year average
Geoje	416 KRW/m^3	391 KRW/m^3	25 KRW/m^3	6.0	20-year average
Goseong	652 KRW/m^3	628 KRW/m^3	24 KRW/m^3	3.7	20-year average
Tongyeong	529 KRW/m^3	495 KRW/m^3	34 KRW/m^3	6.4	20-year average

a. Consignment charge includes operation cost, repayment of investment and commissions given to K-water.

The table above compares the full operational costs of separate systems with the full operational costs of the amalgamated system. The benefits of economies of scale vary. They are driven by such factors as municipalities' access to local water resources, the distance between municipalities involved in the project and the status of services in each municipality. Hence, municipalities considering amalgamation would benefit from examining its feasibility in detail.

Box 2.3. **Amalgamated water services – Gyeongnam Province, Korea** *(continued)*

Comparison of full costs of operation before and after amalgamation

Classification	Unit	Total	Sacheon	Geoje	Tongyeong
Production cost before project (a)	KRW/m³	-	1 139	1 185	1 292
Production cost after project (b)	KRW/m³	-	1 002	1 133	1 102
Difference(c= a-b)	KRW/m³	-	(-)137	(-)52	(-)190
Estimated volumes in sales/year (d)	Million m³	42.2	13.7	16.7	11.8
Total saved cost (e=c×d×f)	billion KRW	118.4	56.4	17.4	44.6
Average saved cost	KRW/m³	50	19	9	22
Contract period (f)	Year		30	20	20

Source: derived from K-water internal reports.

Water tariffs contribute to water resource management

In theory, the price charged water users is an effective tool to promote water-use efficiency, thus contributing to water security when water is scarce. Price-based approaches to water conservation are more cost-efficient than non-price approaches. The gains from using prices as a conservation incentive derive from allowing households to respond to increased water prices in the manner of their choice, rather than by installing a particular technology or reducing a particular use, as prescribed by non-price approaches.

Empirical evidence shows that urban water demand is relatively inelastic to price. In France, a 50% marginal price increase would reduce water demand in a given territory by 10% (see the meta-analysis compiled by Thomas, 2013). In the United States, an average 10% increase in the marginal water price can diminish urban residential demand by about 3-4% (Olmstead and Stavins, 2009). Consequently, pricing alone does not generally suffice to meet urban water policy objectives. While pricing can be effective in the medium to long term, non-price regulation is more effective in the short term.

OECD governments and cities are struggling to design tariff structures promoting wise water use while ensuring service providers' financial sustainability. The OECD (2010a) has shown that few OECD countries reflect water scarcity in their water prices – which instead reflect (at best) the long-term marginal costs of providing WSS services. In Australia, the Productivity Commission has suggested that more research is needed on water scarcity pricing or pricing strategies that reflect different levels of water security or service quality.

Economic theory shows that marginal cost pricing generates revenues that reflect the actual costs of water services (OECD, 2010a); however, it requires heavy data, and reliable projections on water demand and the future costs of supply. Marginal cost pricing can also lead to high prices and expensive water bills, especially when demand is inelastic. Hence, cities in OECD countries often settle for second-best options. The section below discusses recent developments, particularly the social sustainability of sophisticated water tariffs.

Innovative tariff structures

Adapting water tariffs to customer aspirations and providing incentives for efficient water use are important innovations. The best-known examples are the elimination of tariff rebates for large customers (which were based on the idea of economies of scale) and the introduction of progressive tariffs, with either free initial allowances (e.g. in Belgian Flanders) or lower first tiers (e.g. basic allowances are included in the fixed portion of the bill). The objective is to send price signals to customers, while taking their specific situation into account. Similarly, seasonal tariffs use higher prices to cover higher costs during peak demand; they require sophisticated meters that allow reporting consumption immediately after the tariff change. This kind of technology is now available and is no longer expensive.

Budget rates are tailored to individual customers. In Los Angeles, the tariff is progressive (two blocks), seasonal (higher in the three summer months, and also all year-round in a drought year) and adjusted to lot size (large properties are considered as having larger "essential" needs for their gardens and have a larger first block) (Box 2.4). While

Box 2.4. **Innovative tariff structures – California's water budget rate structure (WBRS)**

California faces frequent and prolonged droughts. Water utilities were concerned about the consequences of conservation efforts on their revenues and ability to cover both the fixed and variable costs of service provision. At the same time, customers complained about the equity of existing water tariffs, which did not account for specific household features.

The state experimented with the WBRS to meet four objectives: *(i)* conserve scarce water resources; *(ii)* achieve financial stability of the water utilities even during periods of very limited water consumption; *(iii)* ensure equity and customer satisfaction; and *(iv)* fund conservation and environmental programmes without raising customer taxes.

The WBRS comprises fixed costs and variable costs. The fixed costs are priced at a reasonable level for the customers and the water utility. The variable costs comprise four to six increasing tiers, depending on the water utility. The first tier in each WBRS refers to indoor water use and the second tier to outdoor water use; they represent reasonable use of water by about 75% of the customers. Both tiers are anchored to legal and scientific parameters: indoor water use; the number of residents in the household; the indoor water-use standard per capita; the number of days in the billing cycle; outdoor water use; the evapotranspiration value in inches per acre per day; the landscape factor; the lot size; the drought factor (fraction), representing the water reduction the retail agency is facing; the monthly water allotment; and the days in each month.

The rate structures adjust individual customer tiers based on norms for efficient indoor and outdoor uses. Customers who exceed the first two tiers are considered inefficient and face significantly higher prices per unit of water consumed compared to the second tier. Many water utilities compute the prices of the following tiers by using the next alternative for water (the opportunity cost approach), e.g. imported water or water that is more costly to provide (e.g. desalination of brackish water). Customers can ask for the tiers to be adjusted to their own parameters (variance), which may include the number of people in the household with special needs, the irrigated area, the number of livestock on the premises and the business type.

The revenue collected from higher-tier water use is re-invested in long-term programmes to improve water efficiency and support the water utility urban run-off programmes to reduce aquifer and wetland pollution.

Source: adapted from EPI Water (2011), *Water Budget rate Structure: experience from urban utilities in California,* www.feem-project.net/epiwater/docs/d32-d6-1/CS27_California.pdf.

budget rates are not frequent in Europe, some English water companies propose variable tariffs, with either high fixed portions and low volumetric prices (for customers who have regular water uses) or low fixed portions and high volumetric prices (for people with low consumption, but some peaks).

Providing incentives to use water efficiently can negatively affect service providers' financial sustainability. Cities and water utilities explore options to secure stable revenues even in the face of declining consumption. One option is to decouple revenues from the volume of water sold. California first implemented the water revenue adjustment mechanism (WRAM) and modified cost balancing accounts (MCBAs) in 2008 as part of a pilot programme to promote water conservation.

- The WRAM enables utilities to compensate for any revenue shortfalls from water conservation by authorising customer surcharges.
- The MCBA allows utilities to recoup lost revenue from purchased power, purchased water and pump taxes by adjusting water tariffs to reflect the actual cost of operating the system (Donnelly et al., 2013).

These tariff structures serve multiple goals: *(i)* sever the relationship between volume and revenue, and remove any deterrent to implementing conservation rates and programmes; *(ii)* pass cost savings on to ratepayers; and *(iii)* reduce overall water consumption. The California Public Utilities Commission (CPUC) adopted the mechanisms as part of pilot programmes for conservation rate design. The level of protection enjoyed by utilities using such tariff structures have been the subject of debate.

In an alternative approach, the city of Davis, California, is experimenting with consumption-based fixed-rate water rates (CBFR). The tariff aims to recover all fixed and variable costs, however much water is sold or saved. The rationale for CBFR water rates is that customers pay for the water they use and their share of the system built to bring it to them (Loge, 2013). In this system, water customers pay two fixed rates: the first based on their meter size, the second based on their peak volumetric water use. Revenues generated by these two rates cover the utility's fixed costs. Loge argues that conservation is directly rewarded through lower bills, while rate increases stemming from lost revenue are diluted over the entire ratepayer base and predicated on the individual ratepayer's use of the water system. The new tariff structure will be implemented in 2015, using summer 2014 data as a reference for peak use.

Innovative tariff structures have limitations. Some analysts claim they have contradictory and "disconcerting" effects (Beecher, 2012), with two particularly relevant limitations in the context of this chapter. First (as noted above), water prices have only a limited role in driving water efficiency. Second, pricing policies should pay particular attention to two distinct categories of water users: lower-income users, for whom higher prices raise affordability issues – the next section explores this issue – and wealthier users, who may be particularly unresponsive to price signals, unaware of their water use and less inclined to invest in water-saving technologies and practices (even though they can afford them). Conservation measures could target this group, although such a policy orientation could conflict with financial sustainability objectives.

Social sustainability of water pricing[5]

The financial sustainability of urban water management generates equity and distributional issues (see the discussion above on the water-poor), which are well illustrated in the current debates about increasing block tariffs (IBTs).

The first rationale for introducing volumetric payment of water and IBTs is efficiency water-use and demand management. There is, however, another argument: equity. Arguably, even if price elasticity of demand is small and IBTs have complex consequences, they may still be justified by utilities getting higher revenues from users who generate costly peak demand. Furthermore, on moral grounds, most people believe water wasters should pay: metering and IBTs serve consumer justice.

The trade-offs between efficiency and equity objectives in household water services typically occur when transitioning from an unmetered to a metered charging structure, rebalancing tariffs from fixed charges to volumetric charges, and increasing fees and tariffs to full-cost pricing. OECD countries have considerable experience with policy measures aiming to ensure water affordability for vulnerable groups while passing on the full economic and environmental costs of water services to water pricing (OECD, 2003).

Supporting measures for the poorest families fall into two broad categories: support for targeted household revenues and preferential tariffs. Seen from the perspective of water resource management, social tariffs fail to promote water-wise behaviour. The OECD EPIC survey (OECD, 2014a) establishes that low-income households more frequently engage in water-saving behaviour, but are less likely to invest in water efficiency improvements. Hence, social measures to address water affordability could include measures to facilitate access to water-saving technologies and appliances.

Measures supporting household revenues include social subsidies, vouchers, fractioned payments and debt forgiveness. Many utilities argue that the social dimension of water services should be handled separately or, in the words of the American Water Works Association (AWWA), by "thinking outside the bill" (AWWA, 2005). In collective housing in particular, tenants find it much easier to pay a fixed water charge with their monthly rent than a random variable bill. When they cannot pay, they may need global support for the rent and maintenance charges than for water alone.

One option is to have water, electricity and gas suppliers donate a small percentage of their turnover to a social housing fund, as happens in France (OECD, 2013a). The fund operates at the county level, since county councils oversee social and sanitation affairs. One problem is that this funding can only help people who are temporarily unable to pay; supporting needy people who do not receive bills directly is more difficult. Another option is to identify poor water users and offer them rebates or vouchers, as happens in selected French cities (e.g. through the suburban *Syndicat des eaux d'Ile-de-France*) or in Chile (OECD, 2013a), where authorities are able to identify the people who need this kind of support.

Measures to address affordability issues include preferential tariffs. Designed to keep water bills below a certain fraction (e.g. 3-5 %) of revenue, they entail keeping water charges under a threshold and applying IBTs. Several cities are combining IBTs with social rebates: Dunkirk (France) supplies the first 75m^3/yr block at EUR 0.80 per cubic metre (EUR/m^3) and EUR 0.30/m^3 for families receiving benefits. The second block, up to 200m^3, costs EUR 1.50/m^3, and additional consumption above that threshold costs EUR 2/m^3; there are no social rebates for upper blocks. Since using data on family size and setting the blocks per capita is illegal in France, these figures are multiplied by the number of apartments connected to a meter, irrespective of the number of residents in each apartment. How this social tariff will perform in terms of social redistribution remains to be seen.

Though the OECD (2003) acknowledged that increasing block tariffs may have regressive effects on large poor families, it claimed that "the design of increasing block tariffs can be adjusted in several ways to make the sizes and prices of tariff blocks deliver the intended distributive effects". Some researchers challenge this claim, in particular in the context of developing countries (see Boland and Whittington, 2000; Komives et al., 2005; and a discussion by Baraqué in OECD, 2013a). In practice, where metering is collective and indoor water use is both moderate and inelastic, IBTs may well become a useless complexity. More recently, Whittington et al. (2014) reviewed a variety of subsidised schemes in low and medium-income countries, finding existing subsidies to be very poorly targeted to poor households. They further noted that with each tariff structure under review, "households in richer income quintiles receive a higher proportion of the subsidies than do households in the poorer quintiles".

Experimenting with IBTs requires: *(i)* a general reflection on the distributive effects of tariffs levels and structures; and *(ii)* direct involvement of water utilities in the social dimension of water charges. Now that they have acknowledged the notion of "water-poor" households, cities and utilities must find alternative ways to address the situation.

Diversified revenue streams

The section explores new revenue sources for financing water management in cities in OECD countries. These include innovative business models for water utilities, land-value capture and targeted taxes, e.g. on rainwater run-off.

Innovative business models for water utilities

Some water utilities are searching for additional sources of revenues (e.g. providing "beyond-the-meter" services to domestic and industrial customers) to compensate for declining water consumption. They may benefit from the experience of energy suppliers, who compensate for lower unitary demand with additional value-added services (e.g. stable energy bills, consumption optimisation, or green energy certificates). A growing number of energy providers help customers diagnose their consumption and control their expenses; this additional source of revenue and strengthened relationship with customers may offset the loss in direct energy sales. Electricity companies sometimes offer customers the opportunity to buy "clean" or "eco-friendly" energy at slightly higher prices.

Similarly, water utilities may be well advised to develop policies to help large customers (e.g. hospitals) reduce their water footprint, even though the utility will lose income in the short term. It is always financially wise to anticipate a decrease in consumption rather than suffer the consequences of large customers forsaking the public service altogether (e.g. seaside resorts choosing private desalination).

A few water utilities have set up dedicated teams of water advisers to promote water conservation practices in addition to checking leaky or wasteful appliances. In Los Angeles, a GIS-based software calculates the nominal water consumption of any single-family house; when real consumption significantly departs from nominal value, the utility sends the customer a warning with the water bill, with an invitation to allow its water conservation team to make a visit and help find the source of water wastage. In England and Wales, some water companies have set up new-plumbing departments to provide customers with additional "post-meter" cost-saving services. The utilities derive additional revenues from charging them for the service.

Contractual arrangements can reward utilities for achieving specific performance goals, thus generating additional revenues for them. In Spain, joint public-private companies (*empresas mixtas)* organise the co-operation between local authorities and service providers, and set the conditions for fairly rewarding performance (Box 2.5). In the United Kingdom – where water services are privatised – water companies are concerned that they might lose revenues by supporting conservation measures; the chief executive of Water UK has therefore suggested that they receive tariff bonuses to compensate them for their efforts (OECD, 2013a). Performance-based contracts can set (for example) targets on connection rates, non-revenue water reductions and water or energy conservation. Since experience in the water sector is limited, lessons could be learnt from other sectors – e.g. energy and infrastructure – on the benefits and ways of mainstreaming this type of contract; for instance, in the solid waste industry, performance-based contracts incentivise finding maximum value for recycled goods and increasing the amount of recyclables collected. Lloyd Owen (2013) has identified the emergence of service contracts – particularly performance-based contracts – as the most notable change globally in private-sector water supply and sanitation (WSS) participation.

Box 2.5. An innovative institutional arrangement to manage WSS services – the contract-specific joint public-private company

Several water services are operated by joint public-private companies in Italy, Spain and Czech Republic. This institutional arrangement (called *empresa mixta* in Spain and SEM Contrat in France) works well when the local authorities set explicit and measurable objectives for service quality and when they control the utility which takes stocks in the company to operate the service. The partnership is defined after a competitive process. The utility brings its know-how and expertise.

This arrangement combines four distinctive features:

- The utility is accountable for operational management of the service.
- The company is jointly managed through corporate governance mechanisms set by a shareholder pact. The city mayor or an elected official of the local authority chairs the joint company, whatever the respective shares of capital held by the local authority and the utility.
- The company has a limited duration, aligned with the duration of the management contract. It is terminated when the contract ends, leaving alternative options possible for future arrangements.
- The service is financially sustainable. The operator derives revenues from the service provided (know-how, technologies) and from the eventual return on the equity invested in the company.

The detailed modalities of corporate governance frame the management of the company and determine voting powers, veto rights, etc. They are part of the bid submitted by competing firms. The local authority determines which share of the equity it intends to keep and which is available for the utility. The utility can partner with a financier, for instance when significant investment is required.

This institutional arrangement organises co-operation and consensus between the local authority and the utility, and bypasses the traditional opposition between the contractor and the contractee. Such an institutional organisation can provide a valuable alternative to existing contractual arrangements. Its diffusion in a large number of countries may require legislative adjustments, which would benefit from consultations with local authorities and the private sector.

Source: Semo, I. (2013), "Les conditions de réussite de la SEM Contrat. Le point de vue de l'entreprise", *Actualité Juridique Collectivités Territoriales*, December, pp. 559-561.

Building on concrete illustrations, the OECD (2010b) has developed guidelines for performance-based water utility contracts. The guidelines list the key elements to be considered when preparing, implementing and periodically revising a successful performance-based contracting mechanism. These include: performance indicators; tariff-related issues; contract monitoring; conflict resolution mechanisms; conflict enforcement; and risk mitigation.

Fiscal instruments targeting specific externalities

Taxes can be an effective instrument to address negative externalities that affect water demand and availability, or the costs of water security. Cities in OECD countries have opportunities to consider new taxes to address externalities that have been essentially ignored to date. This section covers two areas:

1. Land development: new developments generate costs for urban water management, for instance when additional reservoirs are required to secure additional demands, or properties are developed in flood-prone areas. Land-value capture can incentivise water-wise land development and generate revenues to cover additional costs.
2. Rain water management: Chapter 1 showed how the built environment can make rainwater management more difficult and costly. Fiscal instruments can help recoup some of these costs and generate incentives for land owners to manage rainwater at the source.

Cities would benefit from systematic monitoring of recent and future developments in the above areas and analysing feasibility conditions. For instance, land-value capture requires establishing a proper regime for land ownership and fiscal capacities.

Land-value capture as a means to finance municipal infrastructure

In a review of financing mechanisms for green urban infrastructures, Merk et al. (2012) note that financing can internalise infrastructure needs, with real estate developers paying to connect their new development to existing infrastructure through development charges (impact fees) and value capture taxes (which capture real estate value increases stemming from nearby new infrastructure development). Developer can, for example, pay development charges (or other financial contributions) to cover the cost of sprawl, thus making it more costly than infill development.

Cities in OECD countries have gained experience in extracting value from land that has benefited from improved services. They have done so either by selling the land, taxing property or the value-added through public investment, or financing infrastructures with the gains generated from land development. They have increasingly focused on land-value capture, with the aim of capturing a share of the increased value accruing from new or improved infrastructure and using it to further fund infrastructure. Successfully conceived and implemented, land-value capture shows interesting possibilities for integrated financial, land use and infrastructure planning.

Cities in OECD countries are also accumulating experience in urban transport systems, particularly roads. Given the strains on the debt financing markets and the fact that user fares are often too low to cover operational expenses (for reasons of social affordability), governments and private operators often need to consider innovative financing instruments to ensure a profitable and sustainable transport infrastructure. Unlike user fares – which capture the direct-use benefit of urban transit – land-value tools capture the indirect and proximity benefits generated by transport infrastructure. They can generate upfront revenues and therefore feature in the capital financing mix, reducing reliance on debt and fiscal risk (Peterson, 2012). Most of the experience in this field has concerned roads, metros and rail. The experience of Casablanca can inspire other cities (Box 2.6).

Box 2.6. **The financial contribution of land development taxes – Casablanca, Morocco**

Casablanca is characterised by rapid urbanisation; its population is expected to grow from 3.5 million to 5 billion by 2030. Extending the water network, securing access to the resource and protecting it against frequent floods are serious concerns for the local authority, which needs to finance these projects.

The city defined a new investment programme in 2007 and contracted Lydec, a subsidiary of Suez Environnement, to provide WSS services and mitigate flood risks. Revenues from user tariffs cover operational and maintenance costs and the renewal of existing assets (accounting for 70% of total cost over the last decade).

A dedicated account (*fonds de travaux*) covers the remaining costs (essentially land acquisition, network extension and social connections). Financed mainly by contributions from property developers, it has financed a growing share of total investment, from 7% in 2004 to 54% in 2014.

Property developers also cover the costs of connecting to the network and in-house equipment. Their contribution varies depending on the type of housing (social housing, villas, hotels and industrial zones), and they pay additional costs for developments that do not feature in the master plan. Contributions are waived when the developments take place in underprivileged neighbourhoods and slums. Special conditions have also been set to adjust the contribution to the pace of urban expansion, and to harness major urban developments.

The contribution is a share of the price of the property when sold, ranging from 0.7% of the selling cost for social housing to 1.3% for luxury apartments and buildings.

Source: personal communication by Suez Environnement.

Land-value capture has its own limitations. In already densely populated and built-up areas, opportunities for land-value capture are limited; relatively undeveloped areas benefiting from new infrastructure have considerably more potential. Peterson (2006) notes that under specific conditions, exchanging landholding for infrastructures can contribute to infrastructure financing. Land leasing can currently only be used as a transitional infrastructure-financing strategy. Yet the supply of land available for lease or sale will eventually run out, and cities will have to rely increasingly on revenues from infrastructure services to recover capital costs. The land leasingfinancing strategy also generates particularly acute risks when real estate prices are highly volatile.

Experience of land-value capture for urban water management is scarce. Faced with frozen local property taxes and restricted local tax increases and municipal borrowing, Californian municipalities turned to land assets as a way of financing infrastructure: newly adopted intergovernmental rules allowed developers to issue land-based bonds to finance roads, sewer and water systems, and other basic infrastructure that could no longer be financed by the public budget (Peterson, 2006). As a result, land became the collateral for much of new infrastructure financing. USEPA has used a similar rationale to mandate property developers to invest in sewerage infrastructures (Box 2.7). In a recent review of water management in the Netherlands, the OECD (2014b) noted that developers who locate the property outside the dykes generate future liabilities and should be accountable for the future costs of protecting their assets.

Box 2.7. Property owners finance sewer renovation – San Francisco Bay area

The EPA has ordered the East Bay Utility District as well as 7 San Francisco Bay communities to spend USD 1.5 billion on upgrading 1 500 miles (2 414 km) of sewerage infrastructure over the next 21 years.

The municipalities will spend an estimated USD 900 million on municipal sewer upgrades; property owners will spend an estimated USD 600 million on private sewer lines by meeting quality requirements applied when properties are sold.

The mandate is interesting on two fronts. First, it confirms that infrastructure upgrade is urgently needed. Second, it points at property owners as a potential source of investment and indicates one way to induce private investment.

Source: adapted from *Global Water Intelligence* (31 July 2014), www.globalwaterintel.com/news.

A tax on impervious surfaces to finance urban drainage

In 2010, France introduced a tax on impervious surfaces to finance urban drainage, with the objective of managing source rainwater before it reaches sewerage networks, thus avoiding run-off and treatment downstream (Box 2.8). Municipalities can set and design this new tax on a voluntary basis. The Great Lyon metropolitan area, for example, offers tax reductions when the landowner takes action to limit run-off (Centre d'études et d'expertise sur les risques, l'environnement, la mobilité et l'environnement [CERTU], 2012).

Box 2.8. Financing urban rainwater management in France

The failure to manage rainwater properly affects the capacity of French local authorities to achieve the "good ecological status" mandated by the European Water Framework Directive. Thanks to a dedicated fiscal instrument introduced in 2011 French local authorities have the capacity to set up a new public service dedicated to urban rainwater management. This new service can be financed in full or in part by earmarked revenues from a dedicated tax.

The tax is based on impervious surfaces, in urban areas or future development areas, whether or not the surfaces are connected to a drainage system. It is paid by the owner of the land or property, when the property is larger than a minimal area set by the local authority. The tax rate is set by the local government and cannot exceed EUR 1 square metre per year (EUR/m^2/year). IT can be reduced, in full or in part, where facilities are in place to reduce run-off; the reduction is meant to reflect the decreased run-off. Several adjacent property owners can join the mechanism, if they build and operate a common facility.

This new tax principally aims to create incentives for managing rainwater close to the source and limiting run-off by implementing measures that mitigate the consequences of impervious surfaces. It also aims to raise revenues, earmarked for long-term urban rainwater management. In the long term, the revenues generated by the tax are bound to decrease as the objectives are met – a trend that local authorities need to anticipate and factor in.

Local authorities have the opportunity when engaging in feasibility studies to reflect on the level of ambition of their urban rainwater management policy and the policy packages (zoning, standards, information, tax, etc.) they wish to implement. Stakeholder consultation should feature prominently in the process.

Source: CERTU (2012), *Taxe pour la gestion des eaux pluviales urbaines*, Ministère de l'Écologie, du Développement durable, des Transports et du Logement, Paris.

In practice, this mechanism is plagued with difficulties. Transaction costs are high, as a large number of bills have to be sent and managed. The main contributor to the tax is the city itself, since it is responsible for the largest sealed surfaces in the metropolitan area – roads. Additionally, the tax can be seen as a transfer from the general budget to water management. Given the small number of large contributors, it may be more efficient to incentivise them through targeted measures. While this may require adjusting the design of the tool somewhat, the rationale is sound and could be explored more systematically.

A role for private investors

This section reviews experience and considers options for using private investment sources, including water service operators, financiers (who do not operate water services) and property developers, to fund urban water management. Private operators' capacity to generate efficiency gains – which translate into reduced financing needs – was acknowledged earlier in this chapter; it is not covered in this section.

Private-sector participation in urban water management

Most OECD countries consider some form of private-sector participation (PSP) as an option to channel additional sources of financing to bridge upfront investment needs. OECD (2009) developed a "Checklist for Public Action" providing governments with a coherent set of policy directions and the necessary steps to engineer a successful partnership with the private sector. The Checklist highlights five key policy areas that can affect public-private co-operation: *(i)* deciding on the nature and modalities of PSP; *(ii)* providing a sound institutional and regulatory environment for infrastructure investment; *(iii)* ensuring public and institutional support for the project and financing choice; *(iv)* ensuring that co-operation between the public and private sectors works in the public interest; and *(v)* encouraging responsible public conduct.

In an overview of water financing options in California, Ajami and Christian-Smith (2013) note that the challenge with PSP is defining how utilities can leverage private capital to invest in projects such as efficiency and conservation projects, system O&M, systematic upgrades and affordability. Private financiers may not find individual projects attractive. Hence, one option is to aggregate projects so that they will be assessed and financed as a single project – an important emerging approach for facilitating investment in small-scale sustainable energy infrastructure. The US company SolarCity, for example, issued over USD 200 million in securities (notes) backed by residential solar photovoltaic leases. Because many developers of sustainable energy projects lack the credit rating required to themselves issue bonds, securitised bonds – where repayment depends on the quality of the assets rather than the creditworthiness of the issuer – have the potential to lower financing costs (Kaminker et al., forthcoming).

Opportunities to access equity finance

As an alternative to PSP, several utilities are exploring options to recycle some of the capital tied up in water infrastructures to generate liquidity for new projects. Long-term investors (such as pension funds) might be interested in contributing and substituting for public money in the utilities' capital structure, provided that the investment's risk-return profile is attractive.

This recent development has far-reaching consequences. First, as noted by Global Water Intelligence ([GWI] April 2014), "the emergence of a relatively liquid market for equity stakes in brownfield water infrastructure projects means that investors who are prepared to assume early risk – including construction risk – increasingly find that there is a natural exit opportunity once a project enters the operational phase." This is particularly the case in a context where the equity market is highly volatile and bond markets only ensure low yields: some water projects typically generate the stable revenues and limited risks that long-term investors long for. Second, this option may be more palatable to public opinion than private investment. It is, however, limited to specific projects, readily isolated from urban water networks.

Typical deals involving private equity firms cover desalination, wastewater treatment and reuse projects for either municipal or industrial clients. Table 2.1 compiles recent major deals where private equity firms invest in water, essentially urban water management. Box 2.9 shares more information about two recent deals in the United States. Such investments, however, are skewed towards specific asset classes: GWI notes that *(i)* they are exclusively located in OECD countries; and *(ii)* they have not exposed investors to single-asset risks.

Table 2.1. **Selected large private equity investments in water**

Private equity firm	Investment	Year of investment	Deal value	Comment
Clayton, Dubilier & Rice	Ashland Water	2014	USD 1.8 billion	Deal pending
Kohlberg Kravis Roberts	South Staffs Water	2013	Undisclosed	Regulated utility
Kohlberg Kravis Roberts	Bayonne Municipal Utilities Authority	2012	USD 150 million	See below
Carlyle	Park Water	2011	USD 102 million	
Kohlberg Kravis Roberts	United Envirotech	2011, 2013	USD 153.8 million	Convertible bond and equity
JPM Asset Management	Southwest Water	2010	USD 427 million	
American Securities LLC	ADS	2010	Undisclosed	

Source: GWI (2014), "Blackstone refreshes its water ambitions", Vol. 15(4).

Box 2.9. **Equity investment in water infrastructure – Bayonne, New Jersey, and Middletown, Pennsylvania**

Bayonne, New Jersey, faces ageing water and wastewater infrastructures and constrained public finance. The local authority intended to develop an arrangement that would eliminate the outstanding debt of the Municipal Utilities Authority (BMUA), advance a sustainable capital-improvement programme and bring long-term stability to the water and wastewater system. Eliminating the debt has improved the bond rating of the City of Bayonne, allowing more efficient use of public money.

In 2012, the BMUA awarded a 40-year water and wastewater concession to a joint venture between United Water and Kohlberg Kravis Roberts. Under the agreement, the joint venture has made an initial payment to the BMUA of USD 150 million, which will be used to eliminate the existing debt and improve the finances of the BMUA. The joint venture has further committed to putting another USD 157 million into the system over the life of the contract. KKR, which made the investment through its infrastructure fund, will fund 90% of the joint venture with United Water.

Box 2.9. **Equity investment in water infrastructure – Bayonne, New Jersey, and Middletown, Pennsylvania** *(continued)*

The BMUA will maintain ownership of the water and wastewater system and will provide oversight of the partnership and its adherence to high quality standards and customer service performance. In addition, t will maintain control of rates charged to users, which will be guided by a formula featured in the agreement. United Water will operate the system for the 40-year term under an O&M agreement with the joint venture.

In theory, the joint venture combines the financial savoir-faire and long-term vision of the financier, and the operational know-how of the private contractor. This investment in Bayonne's infrastructure is expected to improve service reliability and water quality while maintaining rate stability. The funds will be used to upgrade water systems and help ease pressure on municipal balance sheets, freeing the city to invest in other services. The initial capital investment will provide for the addition of highly accurate wireless water metering and other monitoring systems, which help reduce water loss from leakage, prioritise pipe replacement and improve operational efficiency.

A similar arrangement was signed in December 2014 between KKR, United Water and the Borough Authority of Middletown, Pennsylvania, for a 50-year concession contract to extend and operate water and wastewater services.

Under the agreement, the joint venture has made an initial payment to the Middletown Borough Authority of USD 43 million, which will be used to eliminate the Authority's existing debt and pension liability as well as improve its finances. The joint venture has also committed to financing another USD 83 million for the system infrastructure improvements over the life of the contract. KKR, which made the investment through its infrastructure fund, will fund 90% of the joint venture, with United Water funding the remaining 10%.

Source: adapted from multiple sources at the BMUA and United Water.

Several financial techniques can make urban water management more palatable to private investors:

- Maturity transformation consists in taking lots of short-term trades and, using the portfolio effect, turning them into long-term stable demand.[6] This helps mitigate counter-party risk. Such groupings are particularly well-suited to decentralised water sectors, in which small and medium-sized service providers are struggling to access financing on their own merit. In the sector, maturity transformation has mostly been used as a basis for issuing bonds in countries with fairly mature financial markets. High transaction costs and limited knowledge, once again, can partly explain why this technique has remained somewhat limited beyond those markets (OECD, 2010c). Further diffusion may require establishing such grouped financing structures directly (e.g. revolving funds or bond banks) or fostering the adoption of legislation that make such structures more attractive (e.g. tax exemptions on bonds issued by such structures, as practised in the United States, or requirements that grouped financing vehicles be formed in order to access government financing).
- Raising equity can help strengthen the balance sheets of water companies that tend to be under-capitalised. Interesting models have been developed in the water sector (such as the Hyflux Water Trust in Singapore) to mobilise equity through financial markets, thereby diversifying away from mobilising funds from private water companies (whose ability to bring in equity capital is limited in any case) and using such equity injections to leverage other forms of finance for capital investments.

Mobilising equity through capital markets can strengthen financial discipline and improve transparency, including for companies that are primarily government-owned (e.g. a number of publicly listed state water companies in Brazil) (OECD, 2010c).

- Credit ratings can help improve transparency and facilitate borrowers' access to financial markets. Significant progress has been made in awarding credit ratings to municipal governments and water companies, although the use of such ratings has remained limited, particularly in markets that are too small to develop a national rating scale and where the costs of maintaining credit ratings cannot be warranted. The financial crisis has significantly affected the credibility of rating agencies, however, and more generally the reliability of ratings has been questioned in the light of time gaps with regard to information and a potential lack of independence of rating agencies (principal-agency problem) (OECD, 2010c).

The high risk associated with newer technologies can also reduce financing options for innovative urban water management. As discussed in OECD (2013d), risk profiles vary according to the technology and its stage of development, which determines which type of financing, is most appropriate. For example, venture-capital financing is generally suited for unproven and untested technologies, while project finance is used for mature technologies.

WaterTap Ontario is developing and promoting an "Invest to Save" fund for water infrastructure. This initiative is a response to many examples of innovative approaches that cost less than traditional capital projects; however, these projects are typically not eligible for traditional infrastructure funding support. The "Invest to Save" fund aims to support optimisation and efficiency projects that use innovative technologies and approaches in the sector, which will save money for both the province and municipalities and create opportunities for companies to grow while continuing to meet essential levels of service.[7]

Harnessing property developers to invest in water systems

Engineering firms are building water systems using private capital and maintaining ongoing service contracts to finance this capital. Home and landowners are also investing their own capital (or servicing the debt on needed capital) in order to build decentralised systems for single-family or multi-family complexes (Box 2.10). In Mexico, the largest source of investment funding for WSS, besides the federal government, is housing developers (22%). Property developers construct water and sewerage systems within their developments; they have increased their investments substantially as part of large subsidised-housing programmes initiated in 2001 (World Bank, 2005).

Box 2.10. **Innovation in a greenfield site – Brisbane, Australia**

The Payne Road residential subdivision in The Gap, Brisbane, is a greenfield, 20-lot subdivision undertaken by a property developer with an interest in achieving sustainable water management (and who has other projects with similar planned features).

The developer has planned the site to have minimal water transfer in or out. This is achieved through rainwater tanks at each house connected to three large communal tanks, which can be topped up from the town water supply in the rare event of insufficient rainfall. Only backwater is discharged to the existing sewer network; grey water is used for sub-surface irrigation at each property.

Box 2.10. **Innovation in a greenfield site – Brisbane, Australia** *(continued)*

The system architecture combines on-site systems and central infrastructure; this can be consequential for the operation of central wastewater systems. In Australia, water utilities report up to 40% reductions in wastewater collection flows due to, among other things, increased on-site recycling; less water in the system can generate blockages and higher concentrations of contaminants.*

In the Payne Road operation, the developer has kept local and state government stakeholders informed, and these parties maintain an ongoing interest in the project for monitoring purposes. A body corporate will have responsibility for ongoing management of the system's communal components. This project, though small and insignificant in terms of Australia's overall urban water balance, is at the leading edge of decentralised approaches to sustainable urban water management. The water cycle is, to a great extent, localised and "closed loop", resembling much more closely the original natural water cycle than the intervention of conventional centralised systems.

A limitation on replication of this project is the large land areas required (each lot is 1 000 m^2). There are also several unresolved questions such as how water will be supplied during power outages, potential health consequences and social acceptance and amenity over time. In addition, this project does not address the need to close nutrient cycles.

*See the Water Services Association of Australia Report Card 2007-08.

Source: quoted from Livingston, D.J. et al. (2004), *Water Recycling and Decentralized Management: the Policy and Organizational Challenges for Innovative Approaches,* proceedings of WSUD 2004: Cities as Catchments; International Conference on Water Sensitive Urban Design.

Housing/property developers deserve particular attention: in certain contexts, they have incentives to invest in water infrastructure (particularly decentralised systems) to raise their property value. Australia is a case in point: research by the country's biggest property website[8] has revealed that more vendors are seeing green credentials as selling points. Buyers are responding in kind, with 1 in 10 people prepared to pay up to 20% more for a "green" home. As water supplies and sustainability move up the agenda, water-secure properties are becoming more popular; water tanks rank as the feature most likely to add value to a property. In France, public opinion considers rainwater harvesting the second (after renewable energy, and before renewable materials) most positive contribution to green building. However, this appreciation is not reflected in the property value: private homeowners in France have so far failed to recoup their investment cost from their property's resale value.

Innovative institutional arrangements may generate additional incentives for private investment in decentralised systems. In England, "inset appointments" (which allow some customers, particularly large ones, to choose who provides their water supply and sewerage services) generate opportunities to organise decentralised water systems in the context of a central infrastructure (Box 2.11). While they do not refer to either water reuse or wastewater self-treatment, they might set the stage for further developments. Franceys (2007) highlights some of the difficulties associated with inset appointments. The next chapter studies them in more detail, as they are relevant to decentralised ways of providing water.

Box 2.11. Inset appointments in England

In England, inset appointments are an important means of introducing more competition to the water and sewerage industry. Inset appointments were initially allowed for large consumers, typically commercial users such as steel makers and breweries.

In 2007, Ofwat (England's Water Service Regulation Authority) granted Independent Water Networks Limited (IWNL) an inset appointment to supply a 950-home development in Corby, Northamptonshire. IWNL will serve its customers by buying water from Anglian Water and discharging sewerage to Anglian Water's network. IWNL said its 2007-08 volumetric charge will be 5% lower than that of Anglian Water. Ofwat is considering other inset appointments.

Source: Franceys (2007), *Innovative Business Models for Water Supply and Sanitation*, presentation at the OECD expert meeting on Sustainable Financing for Affordable Water Services, Paris, November 2007, www.oecd.org/dataoecd/38/37/40015975.pdf.

The capacity of decentralised water systems to attract (private) investment from parties that will benefit from the rent accrued through improved water services is particularly relevant in a context where public finance is scarce. As Rees et al. (2008) show, given limited government budgets and donor funds, those functions and services that can raise capital or revenue from users or beneficiaries must vitally continue to do so. The opportunity costs of continuing to use public funds to provide private goods to customers who are able to pay for them are high. Remedying this situation requires adjusting the governance structure and financing strategies. Water governance should foster making the best possible use of available water and financial resources.

Similar reasoning applies to other technologies, e.g. sustainable urban drainage. Building on a case study in the United Kingdom, Swan (2010) shows how urban planning controls can incrementally improve stormwater management and reduce run-offs, as well as "transfer some of the incurred costs away from the sewerage provider and onto local developers". As stated earlier, this argues for exploring new fiscal instruments to address externalities related to urban water management.

Notes

1. A later section of this chapter will discuss how two-part tariff structures can address the issue (while raising others).
2. The international community acknowledges 3% as a threshold for the share of the household budget allocated to the water bill. This figure, however, remains subject to debate and could be further disaggregated (by size, revenues, etc.) to reflect the situation of distinct groups.
3. The Netherlands has a distinctive approach, as wastewater is financed by a separate institution (regional water boards) which collects revenues on a large scale.
4. Under the Kyoto Protocol, Non-Annex I countries do not have legally binding emission-reductions targets. They are mostly developing countries. The United Nations Framework Convention on Climate Change emphasises activities that promise to answer the special needs and concerns of vulnerable countries, e.g. investment, insurance and technology transfer.

5. This section relies heavily on a paper drafted by Barraqué (OECD, 2013c).
6. This technique differs from aggregation, which takes lots of small units and turns them into a large-scale parcel.
7. For more information, see www.watertapontario.com/news/blog/invest-to-save-fund-innovative-technologies/28#sthash.1gvVzrMK.dpuf.
8. www.realestate.com.au.

Annex 2.A1

Financing urban water management in Korea

While it has specificities, Korea illustrates some of the trends documented in this chapter.

Water demand and expenditure in Korean cities

Korean household water demand has increased slightly over the last decade. The aggregate figure masks disparities by type of settlements: water demand has increased in mid-size cities,[1] but has remained stable in metropolises and rural areas (see Figure 2.A1.1). In the capital city of Seoul, water demand per capita has declined over the last six years (Figure 2.A1.2).

Figure 2.A1.1. **Household water demand in Korea**

2003-12

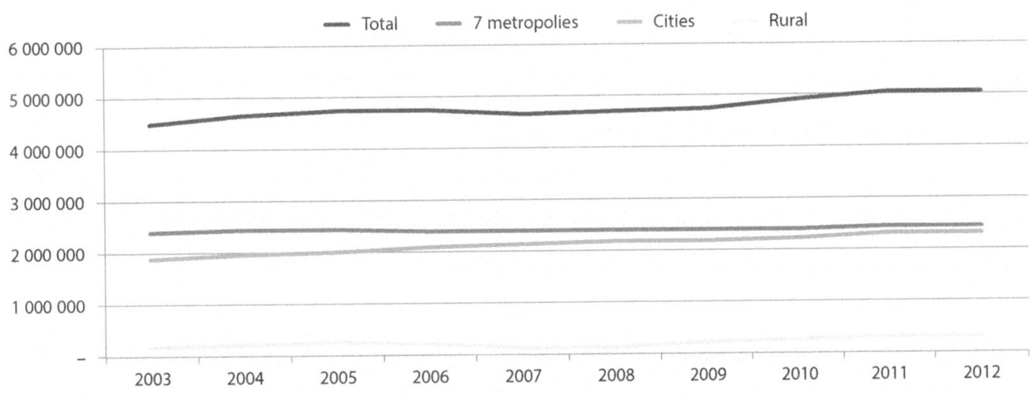

Source: based on K-water statistics (2003-12).

Figure 2.A1.2. **Per capita water use in Korean metropolises**

Litre/consumer/day, 2006-12

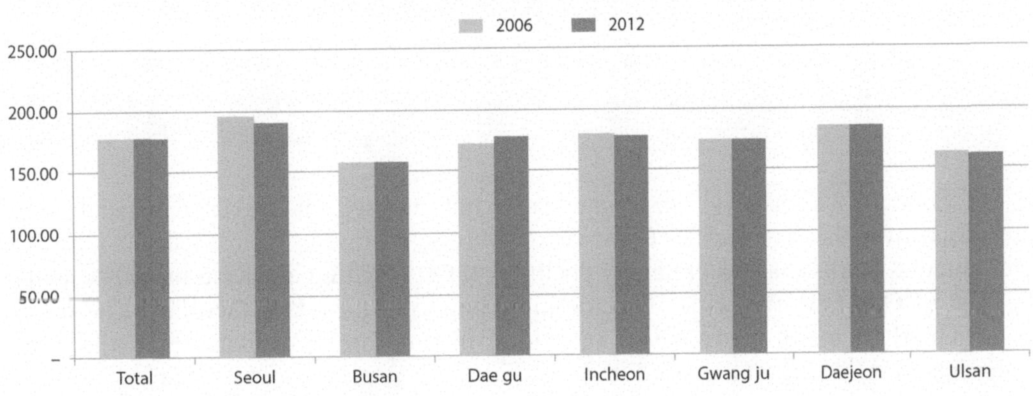

Source: based on K-water statistics (2003-12).

Expenditure on water supply has risen steadily over the last decade (Figure 2.A1.3), both in terms of capital investment (Figure 2.A1.4), and operation and maintenance expenditure. Investment peaked in 2009, driven by a national effort to extend existing infrastructure, especially in cities and urban areas. The share of O&M costs is slightly higher in metropolises, where the larger share of investment goes to improving water infrastructure.

Figure 2.A1.3. **Water supply expenditure**

Breakdown between capital and O&M expenditures, 2004-12 (thousand won)

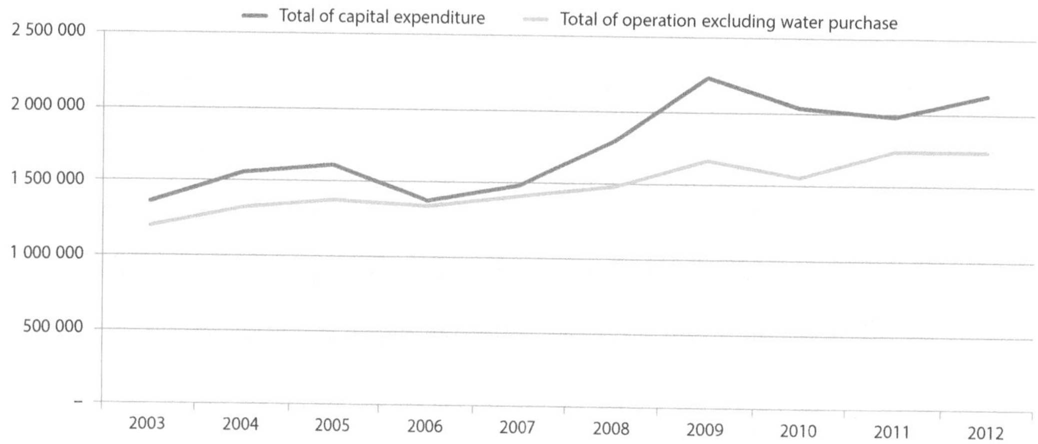

Source: based on K-water statistics (2003-12).

Figure 2.A1.4. **Breakdown of water supply capital expenditure**

2004-12 (thousand won)

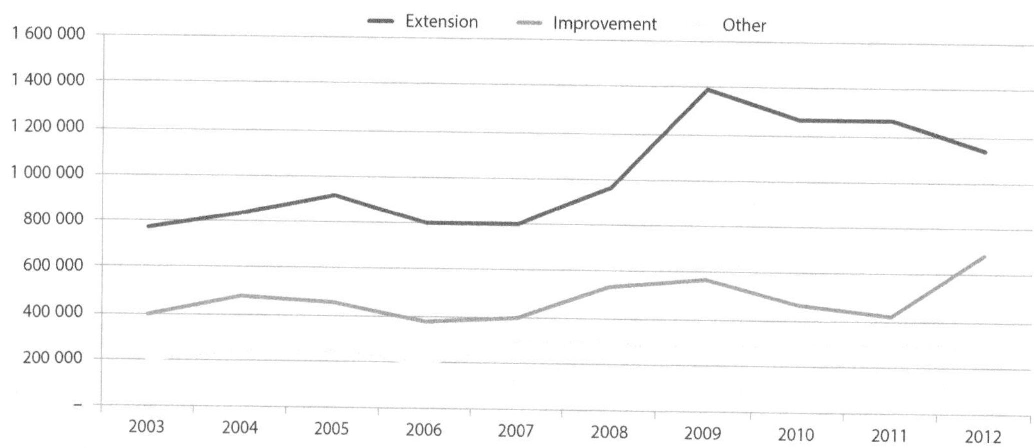

Source: based on K-water statistics (2003-12).

As Figure 2.A1.5 shows, wastewater collection and treatment-related expenditures have risen over the last decade. Capital expenditure[2] represents more than 50% of the total amount, but its share is gradually declining (after reaching a plateau in 2009), while the cost of O&M rises as cities become are equipped with better facilities. The trend is particularly acute in metropolises and cities; in rural areas, capital expenditure still represents over 70% of total expenditures.

Figure 2.A1.5. **Wastewater collection and treatment expenditures in Korea**

Breakdown between capital and O&M expenditures, 2004-12 (thousand won)*

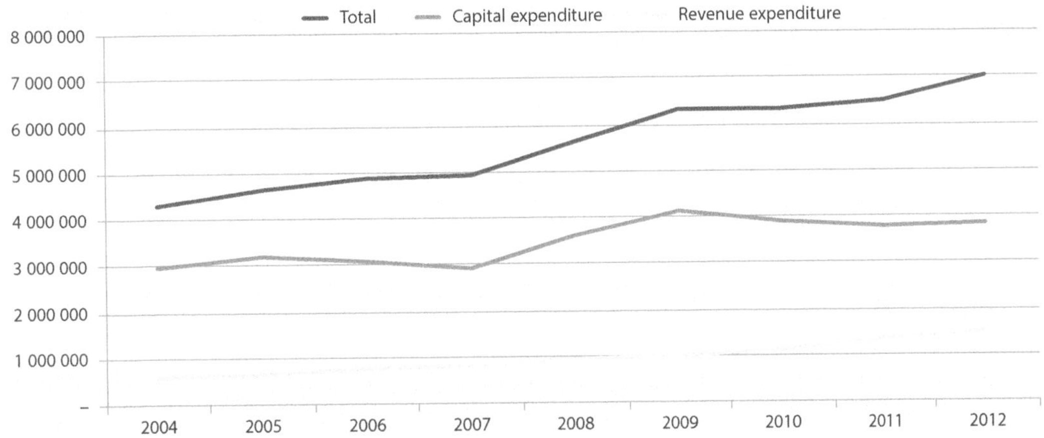

*The figure does not include miscellaneous expenditures.

Source: based on waterworks statistics (2003-12).

Financing WSS in Korean cities

According to the OECD survey on water pricing (OECD, 2010a), Korea stands out as the OECD country where water is cheapest (see Figure 2.4). The situation derives from an atypical combination of financing sources.

Revenues from water supply service bills account for 50% of total revenues, essentially covering O&M expenditure and debt repayment (Figure 2.A1.6). The share of tariffs in total revenues is higher in metropolises (two-thirds on average) and lower in rural areas (about one-quarter).

Revenues from tariffs for wastewater services only account for 20% of total revenues, a share that has remained stable over time (Figure 2.A1.7). Transfers from public budgets

Figure 2.A1.6. **Revenues from tariffs for water supply services**

Compared with O&M and debt repayment, 2003-12 (thousand won)

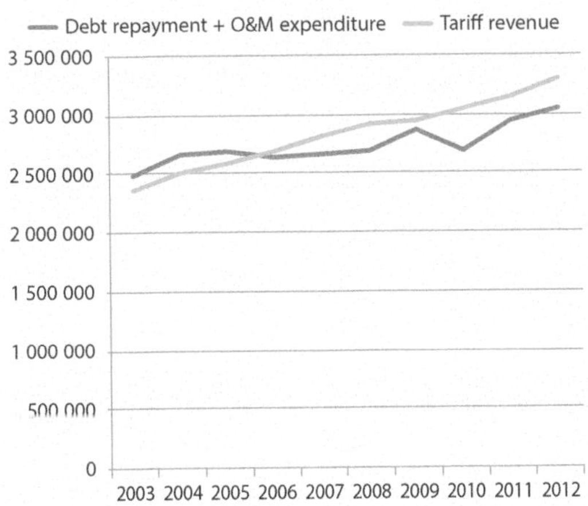

Source: based on K-water statistics (2003-12).

Figure 2.A1.7. **Revenues from wastewater collection and treatment**

Shares of different sources, 2004-12

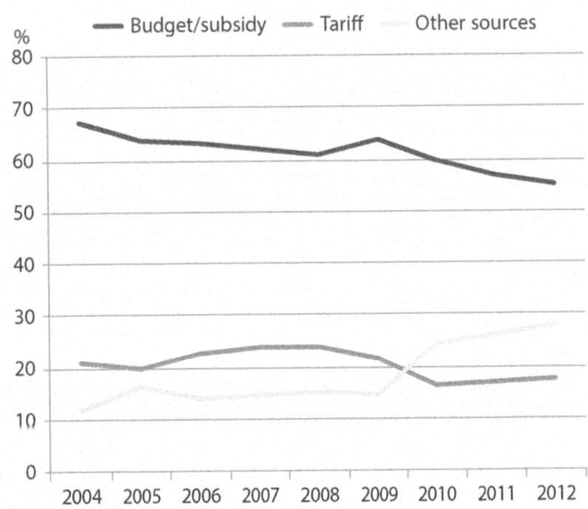

Source: based on K-water statistics (2003-12).

represent more than 50% of total revenues, down from 70% in 2004. The volume of budget transfers has stagnated since 2009. In Korea, an additional source of revenue has gained traction over the last decade: revenues from "others" – including property owners or developers who cover (part of) the costs related to extending the network n. The contribution of these "others" is particularly significant in mid-size cities, where it represented more than one-third of total financing sources in 2014 (compared to less than 10% in rural areas).

Notes

1. Data on water use and expenditure are clustered around three geographical areas: metropolises, which include Seoul and 6 other large cities; cities; and rural areas, defined as *Gun*, i.e. administrative districts with generally fewer than 50 000 inhabitants.

2. Capital expenditure includes extension and improvement.

References

Ajami, N. and J. Christian-Smith (2013), *Beyond Water Pricing. An Overview of Water Financing Options in California*, Pacific Institute, Oakland.

AWWA (2005), *Thinking Outside the Bill: A Utility Manager's Guide to Assisting Low-Income Customers*, American Water Works Association, Denver.

Barraqué, B. (2003), "The Three Ages of Engineering for the Water Industry", Stanford-France STS conference, 7-8 April 2003, http://stanford.edu/dept/france-stanford/Conferences/Risk/Barraque.pdf.

Beecher, J. (2012), "The ironic economics and equity of water budget rates", *AWWA Journal*, http://editiondigital.net/display_article.php?id=964244.

Boland, J.J. and D. Whittington (2000), "The political economy of water tariff design in developing countries: increasing block tariffs versus uniform price with rebate", in Ariel Dinar (ed.), *The Political Economy of Water Pricing Reforms*, Oxford University Press, Oxford, pp. 215-235.

Centre d'études sur les réseaux, les transports, l'urbanisme et les constructions publiques (2012), *Taxe pour la gestion des eaux pluviales urbaines*, Ministère de l'Écologie, du Développement durable, des Transports et du Logement, Paris.

Donnelly, K. and J. Christian-Smith (2013), *An Overview of the "New Normal" and Water Rate Basics*, Pacific Institute, Oakland.

Donnelly K., J. Christian-Smith, H. Cooley (2013), *Pricing Practices in the Electricity Sector to Promote Conservation and Efficiency: Lessons for the Water Sector*, Pacific Institute, Oakland.

European Environment Agency (2013), *Assessment of Cost Recovery through Water Pricing*, Technical report no. 16/2013, European Environment Agency, Luxembourg.

Franceys, R. (2007), *Innovative Business Models for Water Supply and Sanitation*, presentation at the OECD expert meeting on Sustainable Financing for Affordable Water Services, Paris, November 2007, available at www.oecd.org/dataoecd/38/37/40015975.pdf.

GWI (2014), *Blackstone refreshes its water ambitions*, Vol.(4), Global Water Intelligence, Oxford.

Kaminker, C. et al. (forthcoming), *Mobilising institutional investment for sustainable energy infrastructure: defining channels and approaches*, OECD Publishing, Paris.

Livingston, D.J. et al. (2004), *Water Recycling and Decentralized Management: the Policy and Organizational Challenges for Innovative Approaches,* proceedings of WSUD 2004: Cities as Catchments; International Conference on Water Sensitive Urban Design.

Lloyd Owen, D. (2011), "Infrastructure Needs for the Water Sector", unpublished paper commissioned by the OECD in preparation of the OECD report *Water Security for Better Lives*, OECD Publishing, Paris.

Lloyd Owen, D. (2013), "One billion people can't be wrong", *Global Water Intelligence*, November.

Loge, F. (2013), "Consumption-based fixed water rates: a novel conservation-oriented rate structure", *Policy Briefs*, University of California at Davis, Center for Water-Energy Efficiency.

Marsden Jacob Associates (2006), *Securing Australia's Urban Water Supplies: Opportunities and Impediments. A discussion paper prepared for the Department of the Prime Minister and Cabinet*, www.environment.gov.au.

Merk, O. et al. (2012), "Financing Green Urban Infrastructure", *OECD Regional Development Working Papers*, No. 2012/10, OECD Publishing, Paris, http://dx.doi.org/10.1787/5k92p0c6j6r0-en.

OECD (forthcoming a), *Water Governance in Cities*, OECD Publishing, Paris.

OECD (forthcoming b), *Stakeholder Engagement for Inclusive Water Governance*, OECD Publishing, Paris.

OECD (2014a), *Greening Household Behaviour: Overview from the 2011 Survey, revised edition*, OECD Publishing, Paris, http://dx.doi.org/10.1787/9789264214651-en.

OECD (2014b), *Water Governance in the Netherlands. Fit for the future?*, OECD Publishing, Paris. http://dx.doi.org/10.1787/9789264102637-en.

OECD (2013a), "New Modes of Water Supply and Sanitation Management and Emerging Business Models", OECD, Paris, www.oecd.org/officialdocuments/publicdisplaydocumentpdf/?cote=ENV/EPOC/WPBWE/RD%282013%297&docLanguage=En.

OECD (2013b), *Water and Climate Change Adaptation. Policies to Navigate Uncharted Waters*, OECD Publishing, Paris, http://dx.doi.org/10.1787/9789264200449-en.

OECD (2013c), *Water Security for Better Lives,* OECD Publishing, Paris, http://dx.doi.org/10.1787/9789264202405-en.

OECD (2013d), *Cities and Green Growth*, OECD Publishing, Paris, http://dx.doi.org/10.1787/9789264195325-en.

OECD (2012), *A Framework for Financing Water Resources Management*, OECD Publishing, Paris, http://dx.doi.org/10.1787/9789264179820-en.

OECD (2010a), *Pricing Water Resources and Water and Sanitation Services*, OECD Publishing, Paris, http://dx.doi.org/10.1787/9789264083608-en.

OECD (2010b), *Guidelines for performance-based contracts between utilities and municipalities. Lessons learnt from Eastern Europe, Caucasus and Central Asia, Almaty, Kazakhstan*, OECD, Paris, www.oecd.org/env/outreach/47425194.pdf.

OECD (2010c), *Innovative Financing Mechanisms for the Water Sector*, OECD Publishing, Paris, http://dx.doi.org/10.1787/9789264083660-en.

OECD (2009), *Private Sector Participation in Water Infrastructure: OECD Checklist for Public Action*, OECD Publishing, Paris, http://dx.doi.org/10.1787/9789264120525-13-en.

OECD (2006), *Infrastructure to 2030. Telecom, Land Transport, Water and Electricity*, OECD Publishing, Paris, http://dx.doi.org/10.1787/9789264023994-en.

OECD (2003), *Social Issues in the Provision and Pricing of Water Services*, OECD Publishing, Paris, http://dx.doi.org/10.1787/9789264099890-en.

Olmstead, S. and R. Stavins (2009), "Comparing price and non-price approaches to urban water conservation", *Water Resource Research*, http://onlinelibrary.wiley.com/doi/10.1029/2008WR007227/epdf.

ONEMA (2012), *Observatoire des services publics d'eau et d'assainissement. Panorama des services et de leur performance*, Office national de l'eau et des milieux aquatiques, Paris.

Peterson G.E. (2006), "Land Leasing and Land Sale as an Infrastructure-Financing Option", *World Bank Policy research Working Paper*, No. 4043, World Bank, Washington, DC.

Peterson, G.E. (2012), "Unlocking Land Values to Finance Urban Infrastructure", *Trends and Policy Options* no. 7, World Bank Public-Private Infrastructure Advisory Facility, Washington, DC.

Poquet, G. and B. Maresca (2006), *La consommation d'eau baisse dans les grandes villes européennes*, Crédoc, 2006.

PriceWaterhouseCoopers (2013), "Quels mécanismes de financement pour les gares urbaines?", study conducted for La Fabrique de La Cité, Paris.

Rees, J.A., J. Winpenny and A.W. Hall (2008), *Water Financing and Governance*, TEC Background papers, No. 12, Global Water Partnership, www.gwpforum.org.

Swan, A. (2010), How increased urbanisation has induced flooding problems in the UK/a lesson for African cities?, *Physics and Chemistry of the Earth*, Vol. 35, pp.643-647.

Thomas, A. (2013), *Les politiques de gestion quantitative de l'eau*, presentation made at the workshop organised by the French Ministry of the Environment and the Toulouse School of Economy, 11 January.

USEPA (2013a), *Blueprint for Integrating Technology Innovation into the National Water Program*, US Environmental Protection Agency, www.water.epa.gov/blueprint.cfm.

Walton B. (2012), America's water infrastructure shows its age – the national debate about how to pay for repairs, *Circle of Blue*, 5 March.

Water UK (2011), *Sustainability Indicators 2010-11*, London.

Whittington D. et al. (2014), "A diagnostic tool for estimating the incidence of subsidies delivered by water utilities in low- and medium-income countries, with illustrative calculations", *Environment for Development*, Discussion Paper Series, November.

World Economic Forum (2014), *Strategic Infrastructure. Steps to Operate and Maintain Infrastructure Efficiently and Effectively*, World Economic Forum, Geneva.

World Bank (2005), *The Mexico Infrastructure Public Expenditure Review*, World Bank, Washington, DC.

Chapter 3

Supporting the diffusion of innovative pathways for urban water management

The chapter outlines the potential benefits of innovative approaches to urban water management, as well as barriers to their implementation. It covers both technical and non-technical innovation and shows that innovation does not need to be high-tech. It pays particular attention to green infrastructures, smart water systems, distributed infrastructure and urban planning: innovation in these areas can contribute to urban water security, with appropriate business models, at least cost for the communities.

The chapter builds on selected case studies from cities in OECD countries that have embarked on innovative approaches to managing urban waters. It also expands on previous OECD work on policies to support environmental innovation, particularly water management.

Key messages

Innovation in urban water management can be technical (e.g. combining information and communication technology, and water infrastructure) and non-technical as well (e.g. water-sensitive urban planning). This distinction may be elusive, as innovative water management combines technological elements with innovative business models for service providers (whether public or private) and innovative governance (e.g. stakeholder engagement).

Both technical and non-technical innovationch is is burgeoning in cities, and can lower the costs of enhanced water security. However, a series of factors continue to limit its diffusion:

- Retrofitting is difficult, particularly in high-density areas.
- The lack of coherent policies hinders the competitiveness of innovative solutions, e.g. when water prices fail to reflect the opportunity costs of resource use, or when land use and urban development do not reflect the risks of building in flood plains.
- Regulations, funding mechanisms and lock-in failures favour grey infrastructures and incumbent urban water management practices over long-term sustainable practices. They often fail to recognise the capacity of users and the wider communbity to discuss the pros and cons of alternative technologies.
- The lack of data (on river flows or the track record of green infrastructures) weakens the case for innovative technologies.
- Innovative practices may combine different scales in urban water management (from buildings to municipal and larger levels), but can be hampered by institutional arrangements, which split incentives and responsibilities along the water cycle.

Cities that have overcome these barriers have usually combined several initiatives:

- a long-term vision of water challenges and opportunities for urban development
- business models for water utilities and land development that factor in water security externalities
- governance structures that co-ordinate urban water management with other dimensions of urban management and reach beyond city limits
- information campaigns to raise city dwellers' awareness of water-related risks and the liability costs resulting from short-term visions.

Chapter 4 further explores innovative arrangements supporting co-operation between cities and their rural surroundings, and contributing to enhanced water security at least cost to the community.

Introduction

The application of membrane technologies[1] at a decentralised scale, e.g. in households, shopping centres and universities, could completely change urban water systems and open up possibilities for large-scale decentralised urban recycling (van der Steen, 2011). Similarly, the relationship between urban water management and energy consumption could change radically as innovation on waste-to-energy technologies generates energy from wastewater either directly, by using the heat of sewerage, or indirectly, by using nutrient and sludge (for example, the Dutch regional water authorities transform some of the wastewater treament plants they manage in "energy factories"[2]). In the near future, wastewater treatment plants will have the capacity to sell energy to the power grid.

These observations illustrate the potential of disruptive innovation[3] to promote new approaches to designing and operating urban water management systems. They also suggest a pragmatic approach to deploying innovation in urban water management as opportunities arise through the constant renovation of cities in OECD countries. This chapter reviews a selection of such innovations and discusses their potential benefits for cities.

While (hard and soft) technologies are readily available, several barriers hinder their diffusion in cities in OECD countries. The chapter analyses these barriers and explores options to overcome them, using selected innovations – smart water systems; decentralised technologies and distributed systems; fit-for-purpose water supply; green infrastructures; and urban planning – as illustrations.

Opportunities for disruptive technical innovations

This section explores selected disruptive innovations, based on recent developments in cities in OECD countries. It focuses on smart water systems, which exploit advances in information and communication technologies (ICTs); decentralised technologies; and distributed systems, which combine different scales of urban water management.

It bears nothing that no technological option is universally superior: each option generates trade-offs. Makropoulos and Butler (2010) discuss the potential benefit of exploring a portfolio of technological infrastructure options and assessing their benefits in specific contexts. They state that any technology and innovation generates "trade-offs between water use, energy use and land use, and these have an equilibrium point that is associated with the technological state-of-art. At a given technological state-of-art, further reductions in water savings signify increase either energy consumption (for high-tech solutions) or land use (for low-tech solutions). The strategies' evaluation indicates however, that until this equilibrium point is reached there can be significant gains in all three aspects [water supply, wastewater, and drainage]. After this equilibrium, improvements in one aspect inevitably signify costs in others. The choice of desired trade-off then depends on the specific constraints of the problem at hand".

Smart water systems

The International Telecommunication Union (ITU) defines smart water management in cities as the attempt to alleviate challenges in urban water management by incorporating ICT products, solutions and systems in water management and sanitation (ITU, 2014). More specifically, Lloyd Owen defines smart water systems as "systems, components and software that allow the user to monitor, manage and act on data relating to the part of the water cycle that is pertinent to their interests. [They] can be characterised by features including: a high degree of automation, rapid response times or the capability to capture information in real time; the ability to transmit data between remote locations and the data processing facility; and for the data to be interpreted and presented to utilities and end users" (OECD, 2012b).

SWM is data-intensive. In addition to data on water flows and water systems, it often relies on complex mathematical modelling to derive a better understanding of river flows and improve infrastructure design (a point made by Green in OECD, 2013a). The Hamburg case study illustrates how a city can invest in data collection and modelling to respond to emerging challenges; it also points to other areas for innovative urban water management, developed in the subsequent sections.

Box 3.1. **Innovation in urban water management – Hamburg, Germany**

The municipal company Hamburg Wasser (HW) is the main actor dealing with water supply and wastewater disposal in Hamburg. The city's drinking water supply is sourced from the local aquifer; 99% of its population is connected to wastewater disposal and adequate treatment.

HW will soon have to address a few new challenges, such as: *(i)* dealing with an increasing sealing rate, which may lead to overloaded sewers and watercourses; *(ii)* increasing nutrient recovery from wastewater to face future nutrient demand and comply with ecological standards; *(iii)* renovating the ageing infrastructure and adapting the existing infrastructure; *(iv)* protecting against groundwater well salinisation; and *(v)* finding new ways to supply energy and attain high rates of self-sufficiency for wastewater treatment processes.

Hamburg has two main objectives for future water management, namely: *(i)* create a good database of rainfall events and decentralise rainwater retention to avoid urban flooding; and *(ii)* explore different uses of wastewater, e.g. as a source of energy, and recovery of nutrients from waste to be used in agriculture.

The city has taken concrete measures to respond to these challenges and fulfil its objectives.

SYNOPSE

In 2013, Hamburg launched the joint research project SYNOPSE to produce reliable planning instruments for measuring rainfall and minimising future costs. SYNOPSE creates rainfall series to make realistic modelling and planning possible. It involves developing different precipitation models and comparing the resultant digital series with real rainfall data.

Rain InfraStructure Adaptation (RISA)

The RISA project aims to promote sustainable rainwater management and set binding guidelines within a structural rainwater management plan. Its goal is to manage rainwater so as to avoid flooding of basements, streets, or properties, as well as avoid water pollution from combined sewer overflow and urban or street run-off. The background to the project is the rapid expansion of impervious surfaces – by 1 million square metres (m^2) (100 hectares) a year – within Hamburg.

HAMBURG WATER cycle in Jenfeld

The HAMBURG WATER Cycle is an innovative wastewater concept based on source separation, i.e. separation of wastewater streams and energy recovery from the wastewater. As different wastewater streams (black water, grey water and rainwater) have different characteristics, wastewater utilisation is adapted to the specific properties of the material flows.

HW has taken on the responsibility of developing new business models and promoting innovation in the water sector, though this requires time, effort, and strong and loyal partnerships.

Source : the case study was developed by HW for this OECD project.

SMW offers multifaceted potential benefits: it can enhance water quality and reliability, ensure proper management of green infrastructures, decrease water losses due to leakage, reduce operational costs, and improve customer control and choice (ITU, 2014). Particularly relevant from a water security perspective is that smart technologies "can allow water, wastewater and stormwater information to be abstracted and integrated with other data sources such as climate analysis and weather intelligence facilitating a holistic managerial approach to overcome the pressures and challenges within the system". The ITU (2014) further notes that these improvements contribute to both urban wate rmanagement efficiency and financial sustainability, as municipalities and water utilities are better able to recover costs from non-revenue water (e.g. stemming from leakages and illegal connections).

The smart water market is both booming (in terms of the number of players and their market share) and fragmented (in terms of the lack of cross-sectoral applications). To date, companies have tended to concentrate on specific applications (OECD, 2012). For this reason, market trends are difficult to document. Lloyd Owen estimates that global sales for smart water systems ranged from USD 500 000 to USD 1 billion in 2009-10; they are forecasted to rise to between USD 5 billion and USD 16 billion by 2020. Smart water systems accounted for approximately 0.5%-0.9% of the global water hardware market in 2010 and look set to account for 2.9%-9.4% of the market by 2020 (OECD, 2012).

Zooming in: How smart meters affect urban water management

Although smart meters are only one of the typical applications of smart technologies for urban water management, they attract a lot of attention. Initially installed in American cities, smart meters are equipped with real-time information transfer to the utility. They are not mere technical devices: they partake in participatory water governance. They support a new, pro-active relationship with water users, allowing them to have a say in defining the service and better controlling the way they use water. In Boston, meters registering unusual water consumption automatically warn the operator, who can in turn call the customer to check whether this unusual consumption is due to a leak (or another reason). This is possible even though each bulding only has one meter (note that the Boston Water and Sewer Commission has access to crucial information about the number of people behind a collective meter). Smart meters facilitate developing of statistics on per capita water use and analysing water consumption. Box 3.2 presents a case study on the city of Fukuoka, and its use of SWM to improve operational efficiency.

Data collected by smart water meters allow water managers and policy makers to fine-tune the relationship between water consumption and water prices or income. Innovative utilities can build on such data sets to refine their demand forecasts. They can also factor in climate variations, irrigation needs, family sizes, swimming pools, etc. Smart meters can be coupled with "flow trace analysis" software, which breaks down water consumption data by various types of indoor and outdoor uses. This can be very helpful for utilities willing to refine their foresight on water demands, as well as better target inefficient customers and/or appliances that should be replaced.

While smart meters are increasingly being deployed, the analysis of their actual impact on individual water use and customer behaviour is still at an early stage (as reported by Lloyd Owen in OECD, 2012). The technology's full potential has likely not been fully exploited yet.

Box 3.2. Smart water management – Fukuoka, Japan

Fukuoka City had a population of 1.5 million in 2013. The city is not blessed with plentiful water resources. To meet the growing demand for water resulting from its development and population growth, Fukuoka City has constructed an extensive water infrastructure network comprising 8 dams, 5 water purification plants and approximately 4 000 km of water distribution pipelines. However, many of the existing facilities are ageing and need either updating or renewing.

In 1978, Fukuoka City suffered a severe drought caused by abnormally low rainfall and restricted water service for 287 days, devoting a total of 30 000 man-days and a massive amount of expenditures to adjusting for water distribution, typically by opening and closing valves. This operation caused a number of problems, including disruptions to the water supply and in some cases, insufficient water supply.

In 1981, the Fukuoka City Waterworks Bureau developed a water distribution control system aiming to reduce water leakages and promote water-efficient urban planning. It has modified and improved the system repeatedly over the years.

The entire water service area is now divided into 21 blocks. The current system is capable of adjusting water pressure to the proper setting, based on the level of demand in each block. The Water Distribution Control Center monitors and remotely controls a number of water pressure gauges, flow meters and electric valves installed on the water distribution pipes.

The water distribution control system has improved operational efficiency in the following areas:

- Proper adjustment of water pressure within each block has lowered overall water pressure in the entire water distribution system, saving on an estimated 4 000m^3-5 000 m^3 of water leakage every day.
- In the event of damage to the water mains, pipes or purification plants, water supply can be cut off immediately and changes to the water distribution areas made quickly, further minimising water loss from leakage.

Source: the case study was developed by the Fukuoka City Waterworks Bureau for this OECD project.

Policies to support SMW diffusion

In a paper commissioned by the OECD (OECD, 2012), Lloyd Owen reviewed the experience of ten OECD countries in supporting SWM. The analysis highlights how a combination of policies can contribute to diffusing innovation in water management, and notes that innovation can diffuse as an unintended consequence of initiatives in other areas. Lloyd Owen identifies some institutional barriers to diffusing smart water systems. The last section of this chapter will consider these outcomes.

Several countries have explicitly encouraged the development and deployment of smart water systems to improve water management. They tend to focus on smart water metering and the collection and treatment of data to inform consumers on actual water use, manage demand and detect leakages. Smart water meters can be deployed in tandem with a reform of tariff policies (to reflect scarcity), and a series of measures to encourage efficiency. In California and Arizona, policies designed to lower water usage Arizona have led to utilities adopting smart water meters to inform customers about their water usage. In France, incentives to reduce leakage in water supply and sanitation networks have driven the diffusion of smart meters and investment in data monitoring to detect and locate anomalies in real time. Malta is rolling out the world's first national smart water plan, and Jersey is testing a similar approach. In Australia, government funding has supported a smart water meter trial covering an entire community (Wide Bay Water, Queensland).

Other governments (at the national or state level) are promoting smart water systems as part of a policy to support green innovation, as well as information and water technologies. New smart water companies have emerged in Ontario (Canada) and Israel; Korean national policy includes developing a comprehensive network of local and regional smart water grids.

Mixed signals in England and Wales have inhibited smart metering. Given the regulatory structure, utilities engage in five-year spending programmes only; smart metering has too long a payback period for this model. In 2011, Ofwat noted the potential for smart metering, but did not specifically consider it (except for leakage detection and non-revenue water monitoring, where pilot projects are underway). It nevertheless commissioned an evaluation of the costs and benefits of smart metering before setting prices for 2015-20.

Policies not initially targeting deploying smart water systems can have unintended positive outcomes. For instance, competition for non-domestic customers in Scotland since 2008 has triggered the diffusion of smart meters as utilities have striven to improve their performance and customer service. Indeed, advanced metering, pressure management and pipe monitoring systems have reduced leakage with minimal physical intervention in England, Wales and Scotland.

Data-privacy laws in the United Kingdom, Netherlands and California, as well as concerns about the possible health implications of data transmission (e.g. electromagnetic fields generated by radio metre transmission), have prevented installing smart meters in some jurisdictions. Concerns about potential stranded assets can also inhibit their deployment: in Amsterdam, standard water meters were only recently installed, and the utility is reluctant to replace them with smart water meters.

Localised sanitation and drainage at source

Reflecting on future infrastructure needs, the OECD (2006) noted that the most developed countries now recognise that large-scale centralised systems may no longer be viable, due to high maintenance costs and resource needs. This is true for water supply, wastewater infrastructure, rainwater collection and drainage. Recent developments in flood control in the Netherlands point in the same direction (OECD, 2014). This section reviews recent developments in localised sanitation and urban drainage at source.

Localised sanitation

Localised wastewater management systems serve individual or small groups of properties. They can recover nutrients and energy, and can also be connected to local water supply and reuse technologies (Matsui et al., 2001). They require less upfront investment than larger-scale, centrally piped infrastructures and are more effective at coping with the need to expand services (USEPA, 2002). Various commentators suggest that they have a role to play in urban water managvement, even in major developed cities (see for example Tjandraatmadja et al., 2005).

Localised WSS can be used to serve populations not connected to public systems. Rich countries with large metropolises but low population density, e.g. Australia and the United

States, still have significant populations served by private individual or community systems. The situation in Europe is more diverse: the proportion of households not connected to sewers is higher in low-density or low-revenue countries or regions – e.g. Portugal and Spain, southern Italy and Greece, eastern European and Nordic countries, Ireland and even some German *Länder*. In these areas, populations are not yet fully connected to public water systems. Ireland has officially kept a large number of grouped water schemes, providing water to 8% of the population at small community scales (OECD, 2013b).

Localised sanitation systems are not merely a remedy to the limited number of centrally piped systems. They are increasingly used in countries such as the United States, where on-site sanitation now comprises some 40% of all new developments (USEPA, 2002). Sustainable neighbourhoods in cities are partly – or fully – replacing traditional public systems with decentralised technologies. Paradoxically, these innovations take place in the richer and higher-density European Union (EU) Member States (OECD, 2013b).

The performance of localised systems can compare with that of centrally piped infrastructures. For instance, an evaluation of localised systems in Ireland shows that despite difficulties in meeting the standards now imposed at the European level, such schemes sometimes operate better than public water systems, and the population they serve is largely committed to keeping them (Brady and Gray, 2013).

Innovation can contribute to improved performance of localised systems. Research is ongoing to provide communities reliant on individual and community systems with robust and simplified treatment systems, equipped with real-time ICTs, to help set up community services operated from distant centres (e.g. work by Yoram Cohen, UCLA Institute of the Environment and Sustainability).

These developments explain the renewed interest for localised, on-site sanitation. The Australian Academy of Technological Sciences and Engineering (ATSE), for example, recommends that Australian governments encourage investment and uptake of such systems (ATSE, 2012).

Decentralised rainwater collection and drainage

Decentralised systems also apply to stormwater drainage, with a growing use of "source control" technologies that handle stormwater near the point of generation, i.e. locally. For instance, green roofs or pervious surfaces capture rainwater before it runs onto polluted pavements and streets.

These solutions have several merits.

- They alleviate peak flows, by capturing water at source and avoiding run-off onto the streets or in-sewer networks, thereby mitigating the risks of urban floods or sewer overflow in the event of heavy rain.
- They minimise pollution, as rain water gets more heavily polluted when it flows over long distances on dry streets, pavements, or parking lots.
- They improve the quality of water returned to the environment; pervious surfaces in particular allow rainwater to trickle through the ground and recharge aquifers.
- They avoid investing in additional hard infrastructure and treatment facilities: since decentralised drainage limits the flows conveyed through piped infrastructure, it can alleviate the need for extending the sewerage and treatment infrastructure, thereby contributing to cost-effective adaptation to climate change.

- They harness private capital: property and land developers can invest their own money to equip their property with localised drainage systems, particularly in the context of greenfield development; retrofitting existing infrastructures may be more difficult and costly.
- They provide opportunities for direct use (e.g toilet-flushing), since rainwater is collected and available locally, thereby avoiding the costs of transporting water (see Chapter 2 for a discussion of the significance of such costs).

Cities in OECD countries are gaining experience with decentralised rainwater collection and drainage (see the case of Suwon, Korea, in Box 3.3). In Australia, water management at, or near, the source of rainfall through direct roof run-off collection and storage is now part of the portfolio of best practices for providing and maintaining water supplies. Localised rainwater collection can be combined with grey water recycling, and even recycling of sewage water at source.

Box 3.3. **Rainwater Harvesting – Suwon, Korea**

Suwon CIty has a population of over 1.1 million and an area of over 121 km^2. Because it lacks water resources – the ratio of water self-sufficiency is as low as 11% – the city procures most of its water from K-water's multi-regional water supply system.

In 2009, Suwon embarked on the "Rain City" project, designed to reduce dependence on distant water sources and secure enough rainwater to prepare for future water shortage. Water quality was another driver for innovation: non-point sources of pollution continue to flow into streams through rainwater run-off. To control water quality in the urban area, Suwon needed to reduce the volume of rainwater run-off.

The city has taken several actions to promote the Rain City project:

- It has enacted an ordinance (the "Municipal Ordinance on the Management of Water Circulation in Suwon") to establish the legal foundation for promoting rainwater storage facilities.
- It established a master water management programme for 2009 to 2011, which included an analysis of the status of water resources, a plan to improve rainwater collection and use, and guidelines on installing and operating rainwater harvesting systems.
- It installed several facilities for collecting and re-using rainwater, e.g. multi-functional rainwater reservoirs at the Suwon sports complex, rain boxes in private houses and artificial recharge systems.
- It offered to support 90% of the total installation cost of rainwater tanks to encourage citizen participation.
- Finally, it embarked on a public education programme to change perceptions on rainwater collection and use and promote an understanding of how environmental resources circulate in the city.

Building on the project's initial success, the city devised a second plan to be implemented over 2015-18. Endowed with a KRW 10 billion budget, the plan includes installing rainwater-recycling facilities with a 10 000 m^3 capacity and 150 small rainwater tanks.

Suwon's experience confirms that localised rainwater harvesting can be deployed in densely built environments where central piped infrastructure is already in place. However, a comprehensive cost-benefit analysis is required in regions with different climates to assess the economic viability of rainwater harvesting. This assessment should consider water quality, quantity and conservation, environmental impacts, energy saving and the economy.

Source: drafted by Kunwook Kim, K-water, in the context of a secondment to the OECD.

Policy implications

There is increasing experience with implementing and exploiting decentralised sanitation and urban drainage, but remaining barriers (e.g. relating to regulations and business models for water utilities) still need to be overcome. The case study of San Francisco (Box 3.4) illustrates how decentralised water management best materialises when combined with a series of adjustments, including regulation, business models and a new role for the water utility.

Box 3.4. **Decentralised water management – San Francisco, California**

The San Francisco Public Utilities Commission (SFPUC) is responsible for providing retail drinking water and wastewater services to the city of San Francisco. Centralised water and wastewater infrastructure systems have been key to providing high-quality, reliable water and sewer services. As with many other US cities, San Francisco also faces dwindling water supplies, long-lasting droughts and extreme weather events. Most of the options for new water supplies and control strategies tend to be controversial and expensive, driving the city to evaluate new ways of collecting, treating and re-using local water resources.

The SFPUC is embracing decentralised water-treatment systems to provide supplemental water and wastewater services. Large-scale buildings, both commercial and residential, produce alternate water sources (e.g. rainwater, stormwater, grey water and black water) that once treated, can meet their own non-potable water needs. For example, the non-potable indoor water demand for toilet-flushing can be met substantially through capturing, treating and re-using on-site alternate water sources.

However, high costs and the lack of regulatory frameworks can create barriers to implementing the distributed water system. The United States does not apply overarching national quality standards for water for on-site systems using alternate water sources, nor particularly regulations to address ongoing operations and public health concerns.

The SFPUC launched the local "Non-potable Water Program" to regulate on-site water use. The programme creates a streamlined process for new developments to collect, treat and reuse alternate water sources for toilet-flushing, irrigation and other non-potable uses. It also establishes guidelines for developers interested in installing non-potable water systems in buildings, as well as local regulations for ensuring appropriate quality standards.

Subsequently, the SFPUC re-aligned governmental policies and a created new regulatory framework by collaborating with the San Francisco's Department of Building Inspection and the San Francisco Department of Public Health to develop a permitting, review and approval process for on-site system installation and operation. Moreover, the San Francisco Department of Public Health: *(i)* served as programme administrator (providing outreach, technical and financial assistance); *(ii)* carried out cross-connection control (protecting the public water supply, including backflow prevention, testing, certification and tracking); and *(iii)* developed a water-use calculator to help developers estimate the volume of on-site non-potable supplies and demand available for their project.

San Francisco's Non-potable Water Program allowed micro-markets to emerge when two or more buildings share, buy or sell water without the public agency providing the service. The programme shifts the burden of operation, maintenance and quality compliance to the private sector, while the public sector maintains oversight to ensure public health and the protection of the public water system. The move towards smaller on-site water systems holds great promise for reducing freshwater demands, with the aim of building a more resilient and sustainable city.

Source: The case study was developed by Paula Kehoe and Sarah Rhodes (SFPUC) for this OECD project.

Another barrier to deploying decentralised water systems is some communities's perception of being left out when they are not connected to a central infrastructure. Lorrain (personal communication) poitns to a risk of localised systems fragmenting the city into communities benefting from uneven service quality. This potential equity issue needs to be addressed. As Barraqué (OECD, 2013b) notes, the concept of a public service operating non-networked systems is a promising avenue. In France, the 5 million septic tanks currently in operation are now considered technologies that should be kept and upgraded. The implementation of the Urban Wastewater Directive led to a zoning of networked and non-networked areas, the latter being served (or at least controlled) by public services for decentralised sewerage (SPANC). Indeed, the collective management of decentralised technologies creates business opportunities for (public and private) utilities.

More generally, institutions in charge of urban water management would benefit from more systematically exploring the limits of centralised technologies in city peripheriess and the potential benefits of decentralised systems. A fair assessment requires considering urban water management at different scales (i.e. the city and its surroundings, as well as smaller scales where localised systems can be operated) and that urban water governance combine these different scales – a point that the next section will explore further.

Distributed systems

Biggs et al. (2009) define the distributed system model as one where infrastructure and critical services are positioned close to points of demand and resource availability, and linked within exchange networks. Services traditionally provided by a single linear system are instead delivered through a diverse set of smaller systems, tailored to location but able to transfer resources across wider areas. The authors note that distributed systems represent a localised and highly networked approach to production and consumption, and blur the line between centralised and decentralised water models: the central infrastructure plays an arterial role at the regional level, while smaller, tailored systems operate and interact with users at the local level. Distributed water systems combine different operational scales, providing context-specific services while transferring water within and across systems.

Distributed water supply systems have been advocated as part of the "soft path" for water, characterised as an attempt to "improve the overall productivity of water use and deliver water services matched to the needs of end users, rather than seeking sources of new supply" (Gleick, 2002). In a systematic review of cases from Australia, Europe and the United States, Biggs et al. (2009) show how distributed water systems can generate positive outcomes that enhance and supplement those provided by existing infrastructure models. Distributed systems can:

- reduce costs and resource use, by adapting water management to the context and making the most of available resources
- improve service security and reduce the risk of failure, by building redundancies into the system[4]
- adapt to shifting conditions and demands, as well as respond to risk and uncertainty, by increasing the diversity and flexibility of water systems without locking utilities, customers and future governments into rigid pathways for delivering critical services.

The city of Piperton, Tennessee, used sub-surface, interconnected modular wastewater treatment units to service a growing suburban development, linking them within a clustered network that allowed diverting any excess load at one unit to other units as required. Capacity (i.e. additional treatment units) could be added "on-demand" to the network as the population grew, represented a significantly lower upfront cost compared to building or expanding capacity in a centralised treatment plant (Biggs et al., 2009).

In another context, Barraqué notes how distributed systems are adapted to the transition from oversized to more adequate infrastructure (OECD, 2013b). In some German areas, demographic decline combines with decreased per capita consumption induce such a collapse in water demand that public systems end up being largely oversized. Some public operators admit that they will not be able to sustain the present infrastructure; since it would need rebuilding anyway, one option is to redesign them with room for distributed technologies at the single-family, block or community level, particularly in peri-urban areas.

Distributed water systems are not merely technical innovations; they require innovative governance. Porse (2013) discusses trends in stormwater management in future cities. He argues for a hybridised infrastructure combining conveyance and infiltration – which requires hybridised governance, dispersing management and financing responsibilities among central experts and private landowners. He writes that as cities move towards infiltration-based options, they "will likely transfer more responsibility to landowners, such as building and maintaining swales, green roofs, or other treatments on private property". This transition in urban water management governance is consistent with shrinking public budgets and calls for public participation, and echoes developments in Chapters 2 and 5 of this report. However, such governance modes remain unexplored and poorly aligned with the capacities of centralised bureaucracies.

The benefits of non-technical innovations

This section reviews non-technical innovations, which can radically change urban water management in OECD countries, potentially contributing to higher levels of security at least cost for society. It emphasises the capacity to combine a variety of water sources, green infrastructures and sustainable urban planning. Chapter 2 already discussed innovative business models for water utilities.

Combining a variety of water sources

The case for a holistic approach to urban water management, combining WSS with other functions, is gaining traction. It has been a distinctive feature of the global research programme SWITCH, which made a convincing case for a systemic approach to urban water management, stating that the "design and management of the urban water system based on an analysis and optimisation of the entire urban water system (infrastructure and human organisations, water supply, sanitation, stormwater, etc.) will lead to more sustainable solutions than optimisation of separate elements of the system" (van der Steen and Howe, 2009). A holistic approach calls for systems engineering, which designs and operates water bodies (rivers and groundwater) and infrastructures (WSS and stormwater networkks) as one system. The holistic approach also needs to take into account the interfaces between other infrastructures, particularly transport and open spaces. This requires moving away from silo approaches and governance.

Australia's ATSE argues that a holistic approach to urban water sources can be a cost-effective alternative to engineered additional sources of water (ATSE, 2012), and can also be more adaptable to shifts in water availability and demand. In practice, a holistic approach combines decentralised rainwater collection and treatment (as mentioned above), grey water recycling and possibly sewage recycling at source. The case of Tokyo, Japan (Box 3.5) illustrates how the city can use various qualities of water for different purposes.

Box 3.5. **Promoting the use of non-potable water – Tokyo, Japan**

In the early 1960s, Tokyo experienced many droughts, which spurred the Tokyo Metropolitan Government to accelerate the development of water resources. The city renovated its water supply system, expanding it several times to achieve 100% water service coverage. Since the early 1980s, the government has promoted non-potable water use to reduce the burden on sewerage systems and freshwater demand.

In 2003, it established the "Guideline on the Promotion of Efficient Water Usage", designed to provide guidance on using water (including non-potable water and rainwater) efficiently in large-scale construction and development projects. Additionally, as an incentive, the Business Standards Act provides for easing the floor-area ratio restrictions for those buildings installing non-potable water systems to such an extent as to offset the floor space used to accommodate them. Some municipalities provide subsidies for buildings that implement rainwater harvesting, to offset the high costs of installing, maintaining and managing non-potable water systems.

As a result of these efforts, the number of facilities installing the non-potable water system has steadily increased. As of 2012, in-building recycling systems were installed in 408 facilities, 360 facilities had industrial water systems and 1 335 had rainwater harvesting systems.

The Tokyo Dome, Japan's first all-weather multipurpose stadium, is also equipped with both underground storage tanks to collect rainwater from the roofs and a recycling-type non-potable water system that recycles grey water from washbasins and kitchen sinks for non-potable use. These two systems combined provide roughly half of the Tokyo Dome's total demand for water.

The increasing use of non-potable water is not only reducing reliance on surface water, but also contributing to raising public awareness of the importance of saving water. The Tokyo Metropolitan Government continues to move towards comprehensive initiatives promoting efficient water use and measures to prevent leakages, so as to ensure a stable water supply and achieve greater water resource security.

Source: the case study was developed by the Tokyo Metropolitan Government for this OECD project; it is based on data from the Ministry of Land, Infrastructure, Transport and Tourism (2013), *Heisei 25-nenban Nihon no Mizushigen* (Water Resources in Japan).

The holistic approach is supported by emerging concepts – e.g. urban metabolism, which maps water and other flows in a city, thus facilitating combined management (see Fernandez, 2014, for more information and application to cities in OECD countries). Green (OECD, 2013a) notes that most cities export more water than they import, as impervious surfaces convert a high fraction of precipitation into run-off. Even though this run-off is a potential resource that could be used in the city – or downstream – it is in most cases treated as a liability, collected in sewers and transferred to treatment plants before being released into the environment. As discussed above, this approach is costly in terms of infrastructure and generates urban flooding and pollution risks when rainwater is conveyed to sewers, leading to overflow during heavy rains.

While the potential benefits of a holistic approach are better understood, the barriers to its diffusion are also clear. ATSE (2012) notes that greater integration of (natural, recycled or manufactured) water sources in cities will require sophisticated risk management and quality monitoring strategies to guarantee public health. As a range of different qualities and grades of water will be supplied for specific uses, ensuring that the right-quality water reaches the right final user will be critical. The OECD (2009) has argued that regulatory and institutional frameworks tend to pre-empt informed debates on these issues and privilege mainstream, centrally piped management.

ATSE also acknowledges the need for long-term participatory public awareness programmes to overcome negative community perceptions of recycled wastewater and treated stormwater, and facilitate public acceptance of potable recycling. This orientation paves the way for public debate on the pros and cons of alternative techniques and management practices, as well as the level of security deemed appropraite by a community (OECD, 2013c).

Green infrastructures

Green infrastructures are defined as "a strategically planned network of natural and semi-natural areas with other environmental features designed and managed to deliver a wide range of ecosystem services. It incorporates green spaces (or blue if aquatic ecosystems are concerned) and other physical features in terrestrial (including coastal) and marine areas" (European Commission, 2013). They are increasingly recognised as part of the answer to water challenges in OECD countries, especially when cities compete with other users (e.g. agriculture and thermal energy) to access the water they need, and when water management is considered in relation to land use and other policies.

The United Nations Environment Programme (2014) lists green infrastructures for water resource management, some of which are useful in an urban context. Colin Green (OECD, 2013a) adds demand management and local processing of black or grey water to this list. Technologies related to sludge recycling, wastewater-energy generation and water cycle energy efficiency could also be considered (they are listed in Syracuse, an innovative tool for more efficient urban services developed by Suez Environnement[5]). Energy efficiency translates water utilities' objective of minimising and translating costs into opportunities to generate additional revenues (see Chapter 2). Energy-related technologies have ancillary benefits in terms of energy and climate policies. Green infrastructures provide solutions to all four risks that determine urban water security: droughts, floods, pollution and ecosystem resilience.

Most of the technologies inventoried in Table 3.1 are mature. Some have been in use for centuries, e.g. Venice has relied on rainwater harvesting since its infancy and Paris adopted in the 19th century a three-pipe system supplying non-potable water to uses that did not require potable water.

The benefits of green infrastructures are increasingly well-documented. As noted in Chapter 1, the Nature Conservancy (2014) has computed that if cities invested in watershed conservation, 700 million people could receive better-quality water and water utilities could save USD 890 million a year in water-treatment costs. Watershed conservation may be particularlyly relevant to low-income cities that cannot afford the capital and operation and maintenance (O&M) costs of built infrastructures.

The question then is, why are some cities early adopters, while others lag behind? The last section of this chapter will examinebarriers to diffusing green technologies. A point made by UNEP (2014) is worth repeating: green infrastructures work best in combination with existing grey ones, when appropriately sited and designed. The point is not to substitute grey with green, but to combine grey and green, retrofitting green infrastructures to grey ones. This may be particularly challenging in cities already equipped with grey infrastructures and locked-in technical-path dependency.

Table 3.1. **Green Infrastructure solutions for water resource management**

	Urban water management issue							
		Water quality regulation			Moderation of extreme events (floods)			
Green infrastructure solution	WSS (including drought)	water purification	biological control	water temperature control	reverine flood control	urban stormwater runoff	coastal flood (storm) control	Protection of ecosystems
demand management	X							X
local processing of black or grey water	X	X	X					
wetlands restoration/conservation	X	X	X	X	X			X
constructing wetlands	X	X	X	X	X			X
water harvesting	X					X		
green spaces	X	X		X		X		X
permeable pavements	X	X				X		X
green roofs						X		X
protecting/restoring mangroves, coastal marshes, dunes, reefs							X	X
Corresponding grey infrastructure (primary service level)								
dams, groundwater pumping	X			X				
dams, levees				X	X			
water distribution systems	X							
water treatment plant		X	X					
urban stormwater infrastructure						X		
sea walls							X	

Source: adapted from UNEP (2014), *Green Infrastructure Guide for Water Management: Ecosystem-based management approaches for water-related infrastructure projects*, United Nations Environment Programme. and OECD (2013a), *Barriers to, and Incentives for, the Adoption of Green Water Infrastructure*, OECD, Paris.

Water-sensitive urban design

A city's design and build affects water demand and availability, its exposure to risks (particularly floods) and its resilience. Water-sensitive urban design factors in water management: it considers how to enhance water security and access to water services at minimal cost – including financial, social and environmental – to the community.

Water-sensitive urban design starts with a thorough, integratedassessment of water-related risks: it considers water in connection with other domains, e.g. infrastructure, mobility, energy and food. In the Netherlands, new settlements and urban developments must pass a Water Assessment. The assessment plays an important role in co-ordinating water plans on the one hand, and municipal infrastructure and land use plans on the other; it is a process instrument that allows taking into account water management interests (in terms of quality, quantity and safety) in the spatial-planning and decision-making processes. The OECD review of water management in the Netherlands (OECD, 2014) suggested that the Water Assessment could be made more effective (e.g. binding) in influencing these processes.

Water-sensitive urban design is gaining traction as a concept that combines water infrastructure and urban development. As synthesised by van der Brugge and de Graaf

(2010), it encompasses all aspects of urban water management, with additional urban design principles:

- store or use water on site, rather than resort to rapid stormwater conveyance
- capture and use stormwater as an alternative source of water, thus reducing the need for potable water supplied by a utility
- use vegetation for filtering purposes
- use landscapes to protect water-related environmental, recreational and cultural values
- harvest water in decentralised systems for various uses
- treat wastewater in decentralised systems.

Where decentralised and distributed water systems and green infrastructures are concerned, Grant et al. (2012) claim that "distributed water infrastructure can be introduced as part of comprehensive planning strategies that promote compact urban forms with mixed land uses and a focus on urban amenities, encourage alternative forms of transportation to permit narrower streets and reduce demand for parking, foster energy saving and waste recycling, promote water savings, and reduce liability for innovative developers".

Box 3.6. **Water-sensitive urban design – Fukuoka, Japan**

In 2003, Fukuoka City enacted an Ordinance on the Promotion of Water Conservation (hereinafter referred to as "Ordinance") in order to address changes in social conditions and further promote water-efficient urban planning. The Ordinance covers three main areas: the responsibilities of citizens and service providers; the city's responsibilities; and the installation and use of non-potable water systems.

Non-potable water systems fall into the following three categories:

- Area-wide systems: treated wastewater is supplied as recycled water on an area-wide basis for use to flush toilets and water trees; Fukuoka City has operated such systems since 1980 in the context of a programme for recycled water and sewage services.
- In-house systems: wastewater collected within a building is treated and reused as non-potable water in the building; such systems are installed and operated by individual building owners.
- Non-recycling-type systems. rain water, or water from a well, used in the house without prior treament.

Fukuoka City has been offering subsidies to promote the installation of non-potable water systems. For in-house systems, the construction costs for wastewater tanks, wastewater treatment facilities and recycled water tanks receive subsidies equal to 4.2% of the cost of standard facilities. In 2012, such subsidies were granted for a total amount of JPN 940 000 (yen) (EUR 7 000 [euros]).

The Ordinance was intended to build on and further promote water-saving efforts. However, as the volume of non-potable water use is very low relative to potable water, its enactment has had little impact on water tariff revenues.

By promoting water-efficient urban planning, however, the city has been able to save on investments for developing new water resources, such as new dams. Thus, the implementation of the various water conservation measures, including the Ordinance, has had a positive impact on investments in water-related facilities.

Source: the case study was developed by Fukuoka City Waterworks Bureau for this OECD project.

Cities in OECD countries face distinctive challenges in water-sensitive urban design. While developing cities have the opportunity to build right from scratch, most of those that are developed need to retrofit the prevailing design and built environment. The case of Fukuoka, Japan (Box 3.6), illustrates how water-sentitive urban design relates to various forms of innovation. It also shows how diffusing innovative practices can be slow.

In the Netherlands, the Delta programme brings together various authorities and organisations to protect the country against high water and ensure enough freshwater availability. One sub-programme, the Delta programme New Construction and Restructuring, focuses on spatial adaptation and aims to make cities and other built-up areas less vulnerable to extreme weather conditions. In collaboration with civil-society organisations and companies, it entails taking action in three steps: analyse to what extent an area and its functions are water-robust and climate-proof ; propose concrete objectives, and develop effective and efficient strategies, for improving water security and adaptation to climate change; and translate the objectives into policy plans, legislation and regulations, as well as implementation programmes and maintenance procedures. The authorities, as well as the private sector and civil-society organisations can use the specially developed spatial adaptation guidelines developed to work through the steps.[6]

Barriers to the diffusion of innovation for urban water management

Diffusing innovative solutions for water management in cities in OECD countries faces several barriers. Some of them are specific to particular innovative paths discussed above: others are common to all.

Common barriers to disruptive technologies

OECD work on green technologies argues that market mechanisms alone will not provide an appropriate amount of eco-innovation at the right time (OECD, 2011): innovators may not reap all the benefits of their innovations, and markets may not appropriately value environmental benefits. This certainly holds true for innovation in urban water management. For example, when water is not properly priced, markets do not recognise the benefits to water conservation deriving from smart metering or sustainable urban planning; or when city dwellers are unaware of water-related risks, they may be reluctant to pay the costs of higher security levels. OECD (2014) discusses the awareness gap relative to water risks in the Netherlands. In this context, governments are justified in taking action to remedy market failures.

The section below lists three common barriers to disruptive approaches to water management in cities in OECD countries. First, fragmented institutions can limit incentives for diffusing water innovation. Who is in charge of – or accountable for – a particular issue is not always clear, especially when the issue cuts across such domains as urban planning, the environment and economic development. For instance, how should permeable surfaces used for parking slots or streets be defined in the context of urban drainage – as water-related equipments or as transport infrastructures? Who is in charge of managing them? Who is responsible for failures (see the questions raised by Chocat, 2013)?

Second, disruptive innovations have long been perceived as substitutes to existing assets and infrastructures. Indeed, the disordered development of decentralised solutions is likely to undermine the sustainability of centralised systems. City managers now understand better that different pathways need to co-exist and be managed in a co-ordinated manner.

Typically, distributed systems will not phase out centrally piped infrastructures: rather, both systems mutually rely on each other, both to build redundancy and for security reasons.

Biggs et al. (2009) illustrate how new technologies can help identify sites for small-scale system intervention. The retail utility Yarra Valley Water is using spatial-information tools to map the energy intensity of providing water supply and treatment services to different suburbs. This helps identify sites where factors such as gradient, demand and distance from treatment stations mean that small-scale system interventions can save energy. Yarra Valley Water may then choose localised demand management or sewerage treatment as a way to cut costs. This ability to track site-specific resource demand is opening up new areas for mutual cost-cutting between developers and utilities.

Third, regulatory frameworks and funding mechanisms are better aligned with prevailing infrastructures, and sometimes favour or prescribe incumbent technologies (a point made by UNEP, 2014; and Green, in OECD, 2013a). Box 3.7 provides an illustration from the United Kingdom.

Box 3.7. **Retrofitting a sustainable drainage system – Cromer catchment in the United Kingdom**

Swan (2010) explores how extending a city affects the capacity of the drainage infrastructure to collect and treat rainwater. The author compares alternative methods for mitigating flood risks resulting from the expansion of impervious surfaces in Cromer catchment, United Kingdom, as follows.

- Conventional in-sewer storage: a typical conventional solution to flooding in Cromer would involve installing online storage within the sewer network (e.g. in the form of a concrete storage chamber or oversized sewer pipes) to store excess storm flows, releasing them back into the sewer later for subsequent conveyance down to treatment or disposal.
- Retrofitting sustainable drainage systems (SuDS): the most straightforward option in this context involves disconnecting large individual properties from the storm sewer, instead using SuDS devices to deal with their storm drainage. Ground conditions at Cromer make it possible to consider both infiltration and storage-based SuDS devices.
- Hybrid option: this approach involve the combined use of the two preceding options to alleviate catchment-flooding problems.

Swan concludes that SuDS is feasible, appropriate and allows tapping private capital, writing that "the use of SuDS technologies within a 'planning-based' approach, involving the progressive imposition of 'greenfield', or stricter, run-off restrictions to all new planning proposals (both new-build and brownfield redevelopment) submitted within a problem urban catchment may represent a more sustainable way to reduce the storm water run-off entering the system, and the associated problems, over the longer term. This approach would also transfer some of the cost away from the sewerage provider and onto local developers" (Swan, 2010).

Swan's analysis holds that despite their relative merits, SuDS options are difficult to implement because the United Kingdom's current regulatory and funding environment promotes "quick-fix" solutions to urban drainage problems.

Source: Swan (2010), "How increased urbanisation has induced flooding problems in the UK: A lesson for African cities?", *Physics and Chemistry of the Earth*, Vol. 35, pp.643-647.

Specific barriers to distributed system diffusion

Distributed systems face their own limitations and are certainly not appropriate in every context. For example, in case of water shortages, distributed storage forbids allocating available resources where they are most needed. Similarly, in case of floods, piecemeal implementation of sustainable urban drainage can have adverse effects if peak river flows cannot be managed at catchment level in a co-ordinated manner.

Where they are appropriate, distributed systems face several barriers to their diffusion. First, they can weaken existing central systems when the wealthiest consumers disengage from the central network, depriving the managing utility from revenues. This is an issue, as distributed systems work best in combination with centrally piped infrastructures. Utilities and city administrations may be reluctant to explore options that negatively affect their existing networks' revenue base, unless they can identify alternative revenue sources (see the discussion in Chapter 2).

Second, distributed systems raise the issue of liability: who is responsible and accountable for the service provided at the building or district level? In France, the mayor is accountable for the quality of the water supplied to the city's residents. Taking responsibility for a multitude of distributed systems organised at different scales, and supplying water at different quality levels for specific purposes, may be more difficult. Accountability is a challenge because distributed systems require the capacity to monitor and control the quality of multiple water flows at several levels.

Third, distributed systems illustrate the complexity of scale issues. Urban water management generates physical economies of scale, as it is generally cheaper to operate a large treatment plant than several smaller ones – which can be offset by system economies, as treatment and reuse technologies or sustainable urban drainage can save on the capital costs of extending central infrastructures. They may, however, require more energy, as economies of scale apply here as well. Japan – where large areas of the country were not connected to wastewater collection and treatment systems after the Second World War – was able to make extensive use of a variety of on-site recycling techniques.

Biggs et al. (2009) report on research that illustrates the difficulty of defining the right scale for wastewater treament. Researchers compared four distributed wastewater treatment system designs at an urban greenfield site against a centralised reference case. One design, which used 4 treatment plants and on-site greywater reuse, showed how splitting wastewater streams at their source could cut demand for sewerage treatment services by 65%. In another system design, 32 networked treatment plants treated all wastewater and then returned it through a third-pipe return system, reducing reduced residential demand for potable water by one-third and saving 6% on energy compared to the centralised option. The gain in energy efficiency stemmed from the lower energy needs of many smaller treatment plants and the use of recycled water (which reduced energy for pumping). A similar option using 315 smaller wastewater treament plants proved less energy-efficient – suggesting that as the treatment system became increasingly localised (for this site), efficiency increased to a point, and then declined.

This analysis echoes the point, made earlier, that any urban water management technique generates trade-offs between water, energy and land use. It follows that an optimal combination of centralised and distributed systems requires reflecting the externalities related to these three dimensions.

Barriers to diffusing green infrastructures

Green (OECD, 2013a) identifies specific barriers to diffusing green infrastructures for urban water management:

- The greatest problems exist in high-density areas (central business districts), where green infrastructures are most difficult to implement because space is scarce to weave in water management. The existing stock of buildings adds to the complexity, as green infrastructures will require retrofitting. Chapter 2 has argued that in the United Kingdom, sustainable urban drainage accumulates in greenfield projects but has received less attention in existing urban areas (Swan, 2010). Still, Malmö (Sweden), Philadelphia or Portland, among others, increase the level of permeable surfaces and limit run-off.
- As noted above, decision-makers face trade-offs between water, energy and land issues, e.g. when a flood plain is the only land available for urban expansion or economic development, or when water efficiency gains in appliances require additional energy.

These barriers illustrate lock-in failures, defined as "failures where cities are locked into complex infrastructure systems and where the associated skills, knowledge and capabilities are so dominant that shifts to more efficient alternative technologies are difficult to undertake" (Suzenet et al., 2002). Lock-in failures translate into difficulties in adopting new technologies outside of greenfield projects.

Several factors make green infrastructures particularly unappealing for policy makers:

- Data on water is scarce; capacities to measure, compute and model river flows are limited, making it difficult to design and assess the benefits of green infrastructures.
- Uncertainty and the costs of operation, maintenance and safety generate reluctance to adopt innovative approaches.
- While policy makers and city dwellers find it easy to identify large physical interventions as immediate and visible responses to water risks, green infrastructures do not provide the same signal.
- Green lands are not properly valued. In a report for Department for Environment, Food and Rural Affairs, and Communities and Local Government in the United Kingdom, Forest Research (2010) points to good evidence that green space can make positive impacts on local economic regeneration, especially with regard to job creation, business start-up, increased land values and inward investment. However, the quality and quantity of this evidence is comparatively poor and would benefit from further case studies.

The economic case for green infrastructures may be weak, less because of the intrinsic performance of these infrastructures than because of the lack of historical data. The economic analysis of green infrastructures is complex and often lacking track records of past costs and benefits. Ancillary benefits can be numerous, and difficult to assess and monetise; for instance, green infrastructures can mitigate climate change, as they capture carbon and reduce the urban "heat island" effect. As UNEP (2014) notes, green infrastructures can also appreciate in value and function over time as soil and vegetation prosper. This makes assessing their benefits even more difficult, placing them at a disadvantage compared with grey infrastructures.

Overcoming barriers to innovative urban water management

This section analyses strategies that contribute to diffusing innovative approaches to urban water management. It emphasises in particular: *(i)* opportunities for innovative water management stemming from initiatives in other areas – hence the value of opportunistim; *(ii)* the benefit of combining several policy instruments; *(iii)* the need to adjust urban water governance, re-emphasising the interplay between the four pillars identified in Chapter 1 at both central and local levels; and *(iv)* opportunities stemming from the current financial crisis.

Synergies with other policies

Cities can play a crucial role in encouraging innovative water management. The Syracuse project referred to above[7] reviewed selected cities' strategies to deploy innovative water management. The project notes that distributed systems derive either from citizens' initiatives to improve the quality of service where the central system fails, or from demonstration projects in cities wishing to showcase local capacities or to establish themselves as "green" (like Stockholm and Vancouver).

Drawing lessons from the successful implementation of green infrastructures for urban water management, Green (OECD, 2013a) argues that green infrastructures often benefit from initiatives not directly aimed at urban water management:

- River restoration, or the reduction of permeable surfaces, can derive indirectly from climate change adaptation regulations (as occured in London and New York City) or regulations on water pollution from combined sewer overflows (in the United States).
- Incorporating water management into spatial planning and development control systems is a way forward. In Germany, municipalities combine responsibilities for spatial planning and development control for providing water services, including surface water drainage.

This echoes a point made by Lloyd Owen (2006) in the review of country experiences (see the previous section), e.g. that smart water systems may diffuse as an unintended consequence of non-related initiatives (such as opening markets for non-domestic water users in Scotland).

Combining policy instruments

Green (OECD, 2013a) notes that local initiatives resulting in the deployment of innovative urban water management combine several instruments: information campaigns on desired change; demonstration projects (symbols matter); and regulations. In Toronto, regulation was required to expedite downspout disconnections, an essential step in reducing run-off from rainfall and mitigating sewerage overflow risks.

Economic instruments play a specific role. On the one hand, well-designed tariffs can signal resource scarcity and reflect some of the benefits of enhanced water security and improved water services. On the other hand, properly designed targeted and time-bond subsidies to promote innovative water management can be less costly than extending traditional infrastructures. General Electric (2011) reviewed local initiatives to promote water reuse and highlighted the role of rebates and rate reductions. New York City's Comprehensive Water Reuse Programme provides a good example of strategic rate reduction in the context of a uniform volumetric tariff (Box 3.8).

Box 3.8. **New York City's Comprehensive Water Reuse Programme**

New York City's Comprehensive Water Reuse Programme offers a 25% reduction on water and sewer charges for buildings that maintain a Comprehensive Water Reuse System (CWRS). A CWRS building may capture, treat and recycle blackwater (i.e. sanitary wastewater) or greywater (i.e. wastewater from lavatories, showers and clothes washers). The CWRS must achieve a 25% reduction in a building's baseline demand for potable water. Programme rules establish a baseline of 60 gallons per person per day for residential buildings and 10 gallons per employee per day for indoor use in an office building.

Since its inception in 2004, this programme has created an effective indirect subsidy for private water reuse systems. It has been estimated that for a large mixed residential and commercial water user, participation in the programme would reduce operating costs by more than USD 1 million a year by 2012 and close to USD 3 million a year by 2015.

Source: New York City Environmental Protection, *Comprehensive Water Reuse Program*; www.nyc.gov/html/dep/pdf/waterreuse.pdf.

Adjusting urban water governance

As observed above, rolling out innovative water management at city scale requires adjusting urban water governance. Cities will be in a position to explore innovative urban water management more systematically when fragmented institutions are able to co-operate and envision the big picture (water, energy and land use); when cities can collaborate with their surroundings; and when stakeholders can voice their preferences.

The city of Rotterdam clearly illustrates this point. Van der Brugge and de Graaf (2010) argue that the transition towards water-sensitive urban design in Rotterdam was made possible through a combination of policy and institutional innovations. However, city-wide implementation of planned infrastructure requires yet another set of institutional mechanisms related to investments, operation and maintenance: it "either requires additional responsibilities for the existing authorities, or a new kind of authority or co-operation mechanism which receives tasks for operation and maintenance of infrastructure developed by multiple stakeholders" (van der Brugge and de Graaf, 2010).

Opportunities deriving from the current financial crisis

The current financial crisis creates opportunities for low-cost alternative technologies (see Chapter 2). For instance, the US Army Corps of Engineers (USACE)[8] recently noted that "reductions in resources available for construction of federal flood control works present opportunities for expanded implementation of non-structural flood risk management options that are more efficient, less costly, and provide greater environmental benefits. Many of these strategies have been used successfully for years in many parts of the country. They have not always received full consideration, however, because of a historical emphasis on large engineered civil works for flood protection. Today's fiscal realities present the USACE opportunities to collaborate more closely with local communities in providing technical information and other types of support" (National Research Council, 2012).

The USACE strategy echoes several points made earlier about the (implicit) bias towards incumbent technologies, the need to share information on the potential benefits of green infrastructures, and the requirement to co-ordinate local and national initiatives.

Notes

1. Per the European Commission, membrane technologies are used to treat water "by the use of selectively permeable barriers, with pores sized to permit the passage of water molecules, but small enough to retain a wide range of particulate and dissolved compounds, depending on their nature"– European Commission (2010), *Membrane technologies for water applications. Highlights from a selection of European research projects*, Luxembourg, http://ec.europa.eu/research/environment/pdf/membrane-technologies.pdf.
2. http://energiefabriek.com/english.
3. Disruptive innovation, a term coined by Clayton Christensen, describes a process by which a product or service takes root initially in simple applications at the bottom of a market and then moves up market, eventually displacing established competitors. See more at: www.claytonchristensen.com/key-concepts/#sthash.ha6Ofoh0.dpuf.
4. Redundancy is here defined as a "component of a system not strictly necessary but included in case another component fails" (*Concise Oxford Dictionary*).
5. www.safege.com/en/syracuse-2/.
6. www.ruimtelijkeadaptatie.nl.
7. www.safege.com/en/syracuse-2/.
8. USACE is the institution in charge of flood-risk management and water infrastructures.

References

ATSE (2012), *Sustainable Water Management. Securing Australia's Future in a Green Economy*, Australian Research Council, Academy of Technological Sciences and Engineering, Melbourne.

Brady, J. and N.F. Gray (2013), "Analysis of water pricing in Ireland and recommendations towards a more efficient water sector", *Water Policy*, IWA Publishing, Vol. 15, pp. 435-437.

Biggs, C. et al. (2009), "Distributed Water Systems: A networked and localised approach for sustainable water services", *Business Intelligence and Policy Instruments*, Victorian Eco-innovation Lab, University of Melbourne, www.ecoinnovationlab.com/website/wp-content/attachments/234_Distributed-Water-Systems.VEIL.pdf.

Brugge, R. (van der) and R. Graaf (de) (2010), "Linking water policy innovation and urban renewal: the case of Rotterdam, the Netherlands", *Water Policy*, Vol. 12, pp. 381-400.

Chocat, B. (2013), "Une intelligence au cœur de la ville de demain", *Les Cahiers de l'Eau*, Vol. 2, Suez Environnement, November.

European Commission (2013), *Green Infrastructure (GI) – Enhancing Europe's Natural Capital*, COM/2013/0249 final, European Commission.

Fernandez, J.E. (2014), "Urban Metabolism. City Typologies", presentation at the 2014 MIT Europe Conference, Brussels, 21 May 2014.

Forest Research (2010), *Benefits of Green Infrastructure*, report by Forest Research, Farnham.

General Electric (2011), *Incentives for Water Reuse and Recycling*, General Electric Company.

Gleick, P. (2002), "Water management: Soft water paths", *Nature*, Vol. 418, p. 373 (25 July 2002), doi:10.1038/418373a.

Grant, S.B. et al. (2012), "Taking the 'Waste' Out of 'Wastewater' for Human Water Security and Ecosystem Sustainability", *Science*, Vol. 337(6095), pp. 681-686.

I2O (2012), "i2O Water helps Veolia water save 1.5 million litres of water a day", i2O, www.i2owater.com/page/6/?p=zpfeqwqq.

ITU (2014), *Technical Report on Smart Water Management in Cities*, International Telecommunication Union, Geneva.

Lloyd Owen, D. (2011), "Infrastructure Needs for the Water Sector", unpublished paper commissioned by the OECD in preparation for the OECD report *Water Security for Better Lives*, OECD Publishing, Paris.

Makropoulos, C.K. and D. Butler (2010), "Distributed Water Infrastructure for Sustainable Communities", *Water Resources Management*, Vol. 24, pp. 2795-2816.

Matsui, S. et al. (2001), "Emerging Paradigms in Water Supply and Sanitation", in Maksimovic, C. and J.A. Tejada-Guibert (eds.), *Frontiers in Urban Water Management – Deadlock or Hope*, pp. 229-263.

OECD (2014a), *Water Governance in the Netherlands. Fit for the future?*, OECD Publishing, Paris, http://dx.doi.org/10.1787/9789264102637-en.

OECD (2013a), *Barriers to, and Incentives for, the Adoption of Green Water Infrastructure*, OECD, Paris, www.oecd.org/officialdocuments/publicdisplaydocumentpdf/?cote=ENV/EPOC/WPBWE/RD%282013%298&docLanguage=En.

OECD (2013b), *New Modes of Water Supply and Sanitation Management and Emerging Business Models*, OECD, Paris, www.oecd.org/officialdocuments/publicdisplaydocumentpdf/?cote=ENV/EPOC/WPBWE/RD%282013%297&docLanguage=En.

OECD (2013c), *Water Security for Better Lives,* OECD Publishing, Paris, http://dx.doi.org/10.1787/9789264202405-en.

OECD (2012), "Policies to Support Smart Water Systems. Lessons from Countries Experience", unpublished paper by David Lloyd Owen commissioned by the OECD.

OECD (2011), *Better Policies to Support Eco-innovation*, OECD Publishing, Paris, http://dx.doi.org/10.1787/9789264096684-en.

OECD (2009), *Alternative ways of providing water. Emerging options and their policy implications*, OECD, Paris, www.oecd.org/env/resources/42349741.pdf.

OECD (2006), *Infrastructure to 2030. Telecom, Land Transport, Water and Electricity*, OECD Publishing, Paris, http://dx.doi.org/10.1787/9789264023994-en.

Porse, E.C. (2013), "Stormwater Governance and Future Cities", *Water*, Vol. 5, pp. 29-52.

Steen, P. (van der) (2011), *Application of Sustainability Indicators within the Framework of Strategic Planning for Integrated Urban Water Management*, SWITCH Deliverable 1.1.7.

Steen, P. (van der) and C. Howe (2009), "Managing water in the city of the future; strategic planning and science", *Reviews in Environmental Science and Bio/Technology*, Vol. 8, pp. 115-120.

Swan, A. (2010), How increased urbanisation has induced flooding problems in the UK/ a lesson for African cities?, *Physics and Chemistry of the Earth*, Vol. 35, pp. 643-647.

Tjandraatmadja, G. et al. (2005), "Rethinking Urban Water Systems – Revisiting Concepts in Urban Wastewater Collection and Treatment to Ensure Infrastructure Sustainability", *Water Science and Technology,* Vol. 5(2), pp. 145-154.

UNEP (2014), *Green Infrastructure Guide for Water Management: Ecosystem-based management approaches for water-related infrastructure project*s, United Nations Environment Programme.

USEPA (2002), *Onsite Wastewater Treatment Systems Manual,* EPA/625/R-00/008, US Environmental Protection Agency, Washington, DC.

USEPA (2010), *Green Infrastructure Case Studies: Municipal Policies for Managing Stormwater with Green Infrastructure*, EPA Office of Wetlands, Oceans and Watershed, Washington, DC.

Chapter 4

Urban-rural co-operation for water management

As most cities share water basins with rural areas, an efficient and environmentally sensible urban water management system needs to consider the interplay between urban and rural water uses.

This chapter identifies the main challenges at this interface, which will require cities to co-ordinate increasingly with rural water users in the surrounding areas. It highlights the benefits of enhanced co-ordination and discusses some of the options they can explore to realise such benefits.

Key messages

Many cities in OECD countries share water resources with surrounding rural areas and increasingly have to co-ordinate with rural water users to reach their water management objectives. Three specific water challenges in particular – increased competition for water resources (scarcity), flood management (abundance) and freshwater quality conservation (pollution) – may benefit from co-ordination between urban and upstream (rural) communities.

Co-operation allows managing water at the relevant spatial scale: both the drivers of the problems and the consequences of the chosen policies have significant effects across different areas, jurisdictions and economic sectors. It is also a means to allocate risks to the parties best equipped to manage it. For instance, catchment protection programmes rely on the ability of farmers to avoid pollution and costly treatments downstream.

Cities in OECD countries use a range of approaches to co-ordinate water management with their rural surroundings; many of these approaches involve customised mechanisms to engage with parties and address well-defined objectives. Some necessary conditions are required for cities to take action beyond the existing regulatory frameworks. Their role and degree of freedom in the larger setting of water resource management play a critical role, together with the costs and benefits of engaging action with rural areas. Transaction costs of more sophisticated co-operation arrangements may hinder the diffusion of these actions. Hence, incentive mechanisms and key governance principles are essential to the success of any approach, as they contribute to building trust.

Whether, and how, cities should engage with their rural surroundings remain practical questions that require assessing on a case-by-case basis. Flexible and nimble approaches that take into account the underlying drivers of the constraints they address (such as climate variation and population growth) are promising.

Introduction

An efficient and environmentally sensible urban water management system needs to take into consideration the interplay between urban and rural water uses. Most cities share water basins with rural areas with different populations, stresses, infrastructure constraints and activities. Agriculture, in particular, is a major user of land and water – it accounted for 36% of the total claims on national land water and 44% of the claim on national freshwater in OECD countries between 2008 and 2010 (OECD, 2013a) – and hence can play a critical role in sustainable urban water management.[1]

Several issues generate water management interdependencies between cities and their rural surroundings. The OECD survey on urban water governance (OECD, forthcoming a) identifies at least eight important issues calling for rural-urban co-operation (see Figure 4.1). They explicitly refer to three of the four water security challenges presented in Chapter 1:

- scarcity: the projected growth in competition for water resources, linked particularly to rising uncertainties about water supply due to climate change
- abundance: mitigating flood risks, bound to increase in the future
- pollution: maintaining adequate water quality standards, both for drinking water and ecosystem services.

Figure 4.1. **Issues generating interdependencies between cities and surrounding areas**

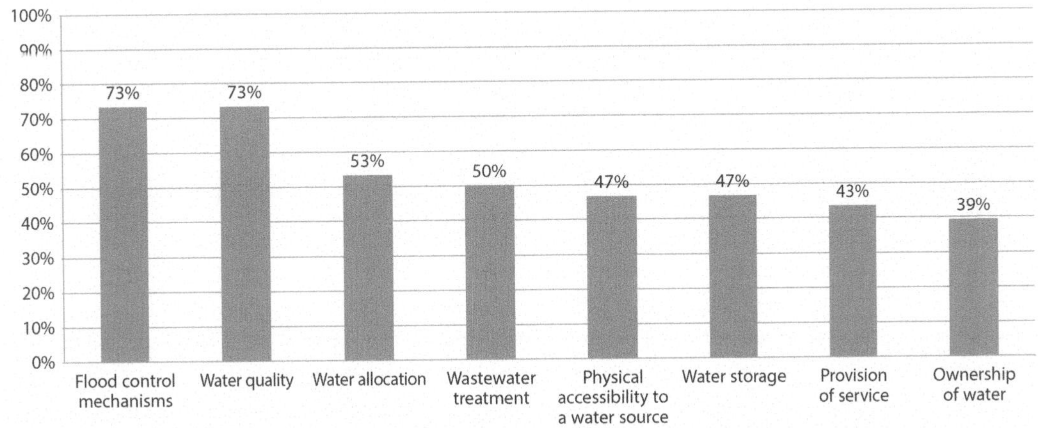

Note: Results based on a sample of 30 respondents who indicated the issues as being "very important" and "important".

Source: OECD (forthcoming a), *Water Governance in Cities*, OECD Publishing, Paris.

Very strong environmental and economic linkages connect urban and rural water users, including flows of agricultural goods, manufactured goods, people, information and ecosystem services between the two areas. Governments could therefore use this strong economic regional interdependence as a platform to address the three water security challenges. Collaboration could be beneficial for all the parties involved: cities can benefit from water savings and quality improvements upstream, and in turn help farmers make the best use of water by incentivising or compensating those who contribute to water security. The United States Environmental Protection Agency (USEPA, 2014) acknowledges that when properly managed, "urban waters can yield positive impacts for populations in both urban and upstream communities: public spaces along rivers and lakes offer residents opportunities for community gatherings, recreation and environmental education [...]; and increased access to waterways can spur the creation of new jobs and the growth of local businesses, whether cleaning up polluted or abandoned properties, opening businesses along the waterfront or working on water protection efforts like green infrastructure projects."

The urban-rural perspective also allows managing water at the relevant spatial scale, since both the problems and the chosen policies have significant consequences across different areas, jurisdictions and economic sectors. Moreover, cities can target their policies in a way that national or regional schemes may not be able to, leading to less costly policy action. A case in point is flood mitigation activities, which are generally interdependent or generate risk externalities; for instance, channelling rivers upstream can increase flood risk downstream, and hydraulic infrastructures and land-cover composition can also affect flood risk in urban areas.

Co-operation can also help manage water risks by spreading and allocating risks among the actors best equipped to mitigate them, possibly increasing response flexibility and adaptability to shocks (for instance, catchment protection programmes rely on farmers' capacities to avoid pollution that would require costly treatment downstream). Cities and their surroundings also share significant uncertainties associated with managing water resources, both quantitatively and qualitatively. In particular, surface water interactions with groundwater, and the diffusion of pollutants at the regional level, are not always perfectly known.

The next three subsections will analyse and frame three water challenges in economic terms: *(i)* competition for scarce water; *(ii)* flood management; and *(iii)* quality preservation (pollution). They do not explicitly cover risks to ecosystem resilience, subsuming them instead in the others. For example, flood control and water storage may rely on flood plains, wetlands or groundwater recharge, and water quality and wastewater treatment affect freshwater system resilience. A fourth subsection will discuss partnerships as a mechanism for implementing rural-urban co-operation.

The analysis focuses on policies and initiatives used by cities and their surrounding areas to discuss the key challenges described above. It does not delve into regional, national and international policies that address these challenges at a higher scale. In practice, national or supra-national quality standards often play a critical role in urban and peri-urban water management as either enablers or barriers to local policy initiatives; however, they will not be considered here if cities are not involved in their design. The measures described here often build on and complement national standards, and/or provide innovative approaches to overcoming specific constraints not covered by overseeing policies and standards.

A second limitation is that the analysis is conducted in a scale-neutral fashion – i.e. it does not factor in differences in city scale, context and characteristics – both for purposes of simplification and to endeavour to reach general conclusions. It acknowledges, however, that these differences will likely affect the feasibility and adaptability – if not the effectiveness – of proposed instruments, and would therefore warrant further analysis.

Managing competition between urban and rural areas to access water

Chapter 1 documented competition between water users and discussed future projections. Competition between cities and agriculture is less acute in OECD countries, where agriculture accounts for 44% of total freshwater withdrawal, compared to 70% globally (OECD, 2013a). Still, where water is in limited supply, and particularly in times of scarcity (droughts), the projected growth in domestic and industrial water demand will create tensions over the current agricultural water allocation. This increased competition may also have external effects, particularly on groundwater (OECD, forthcoming b). Over-drafting the resources can lead to stream depletion (affecting river ways and lakes), salinity intrusion and land subsidence affecting both rural and urban communities. It may also result in depleting groundwater reserves, with long-term consequences such as reduced resilience of agricultural and urban areas. This section discusses in more detail ways of managing which competition over water between urban and rural areas.

Policy objectives and responses

Competition creates a scarcity value for water, which will need to be allocated between the different regions and users in multiple ways (OECD, forthcoming c). All the available policy choices aim to yield the same result: a defined distribution of water abstraction entitlements to specific individual economic agents. Box 4.1 introduces the possible policy responses to the water allocation problem.

Box 4.1. **Water allocation policies**

Given an exogenous variation in water supplies, achieving an efficient allocation of water across users entails redistributing the current water stock. This could be achieved in many different ways, such as administrative transfers, legislative settlements of conflicting claims, legal challenges to existing water allocation, public agency exercise of eminent domain, or the redesign of large-scale water projects to favour different sets of users.

Possible water policies aimed at managing competition for water resources mainly differ with respect to the degree of government intervention and the targeted economic dynamics. This section considers in more detail four possible measures to turn urban-rural competition for water resources into a mutually beneficial endeavour:

- command-and-control instruments: technology standards and rationing policies
- water conservation measures in urban and agricultural areas
- economic instruments: water pricing and tradable water rights
- supply-side groundwater management tools.

Overall, the different water competition policies should not be seen fixed; many sound policies combine and integrate different strategies. Moreover, as previously pointed out, one policy option is not always better than the other and the specific context – i.e. the degree of market formalism in water interactions or level of water scarcity – will have a significant impact on its successful execution.

Sources: Colby (1990), "Enhancing instream flow benefits in an era of water marketing", *Water Resources Research*, Vol. 26, pp. 1113-1120; Olmstead, S. (2010), "The Economic of managing scarce water resources", *Review of Environmental Economics and Policy*, Vol. 4(3), pp. 179-198; OECD (forthcoming c), *Water Resources Allocation: Sharing Risks and Opportunities*, OECD Publishing, Paris.

The resulting water allocations should be assessed using the following three criteria: efficiency, equity and sustainability (OECD, forthcoming c). However, these standards do not always point towards the same direction, which often creates significant political tension. Policy makers will therefore need to strike a balance between these objectives when formulating their water management policies (OECD, forthcoming c).

Efficient allocation maximises efficiency and the total benefits to society produced by water and entails equalising marginal net benefits (MNB) across the allocated users, i.e. the benefit derived from using an additional unit of water should be equal across urban and rural regions. If the available water resources are renewable (e.g. surface water), the current period's flow should be optimally allocated among competing users. If instead the main source of accessible water is non-renewable (e.g. in some groundwater-dependent regions), allocation should be optimised across multiple periods, which entails equalising the discounted MNB to achieve private marginal net benefits across time and users. Box 4.2 describes three idealised scenarios of efficient water allocation between urban and rural consumers (for the case of a renewable resource).

Second, governments should strive for equity in their plans to provide sufficient, safe, acceptable, physically accessible and affordable water to all households, regardless of their income or social status. Policy makers might in some cases favour the equity and fairness of the resulting allocation profiles over their economic efficiency. For example, low-income farmers in water-stressed regions may not receive enough water to sustain their livelihoods; an equitable allocation would therefore require shifting water resources to the more vulnerable groups in society at the expense of economic efficiency. Chapter 5 will further emphasise procedural equity, i.e. participation of all stakeholders in the way water is managed as a whole, especially when it crosses urban and rural areas. New Zealand provides an illustration related to the rights of indigenous people: regional and district plans for managing water in urban and rural areas factor in iwi/Maori perspectives.

A third objective of resource allocation is to achieve a sustainable level of water consumption, which entails both maintaining a functioning hydrologic cycle and preserving the environmental services associated with water ecosystems. Since these uses represent only residual claims on the available water stock, in the absence of regulation they will be satisfied only after both urban and rural water demands have been satisfied. In a context of increasing competition for water resources, government policies need to protect these residual claims in order to avoid major environmental problems, e.g. irreversible damages to freshwater ecosystems or natural imbalances in the natural water cycle. They can do so, for example, by separating the volume of water needed to meet the system's ecological needs and environmental flows from the allocable water available to residential, industrial and agricultural users.

Command-and-control instruments: Technology standards and rationing policies

The first policy option for dealing with the projected growth in competition for water resources is to use prescriptive or command-and-control strategies, of which the two most common examples are technology standards and water rationing policies (Olmstead, 2010). Technology standards force a reduction in water consumption by requiring industrial manufacturers, municipal water providers or farmers to adopt given infrastructure types or production practices. These measures can also be targeted to specific domestic uses of water, e.g. by mandating the use of low-flow appliances. The empirical evidence on the success of these programmes is mixed. In particular, technology standards may be subject

Box 4.2. Economics of rural-urban co-operation

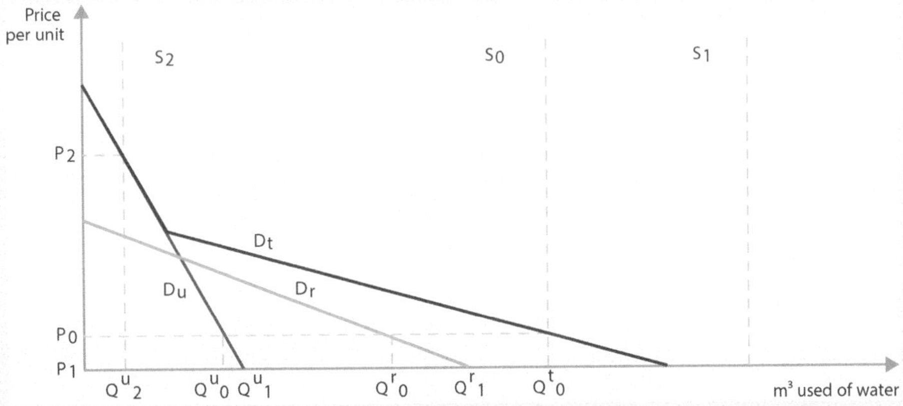

The dark (Du) curve represents water's MNB for urban users and the light (Dr) curve represents its MNB for rural users; these can also be interpreted as their demand schedules. Agricultural demand for water is generally more elastic than urban demand: agricultural producers have relatively more possible substitutes to water use than urban consumers, whose water demand is mainly driven by the fulfilment of basic human needs. For instance, farmers could switch to less water-intensive crops or adopt water-saving production technologies. The black curve (Dt) is the horizontal sum of the MNB and represents the aggregate water demand of the urban and rural regions. The kink in the aggregate demand curve characterises urban and rural residents' different willingness to pay.

The graph depicts three different scenarios depending on the availability of the fixed water supplies: normal conditions (S0), water abundance (S1) and water scarcity (S2). Under normal conditions, efficient allocation is achieved at price P0; total water supply (Q0t) is divided among urban and rural users (Q0u and Q0r), such that their marginal benefits of use are equalised. In case of water abundance, supply exceeds the total demanded quantity; there will be no competition for water resources and therefore the efficient price of water will be P1 (opportunity cost of water). Under severe water scarcity conditions (S2) such as droughts, the optimal strategy to minimise total damages is to impose a larger reduction of the water supply on the more price-elastic user, namely agriculture. Efficient allocation will therefore entail allocating the entire water supply to urban consumers, which will pay the higher price P2; rural regions will therefore not receive any water supplies. From a strictly economic point of view, this allocation is optimal because the opportunity cost of diverting resources from urban users is higher than the corresponding figure for farmers.

In all three scenarios, efficient allocation is achieved through market interactions among competing water users. The market outcome is a set of water allocations and the corresponding price of the market transactions. This is, however, a highly stylised representation of the problem that does not take into account the complexities of the allocation process, transaction costs, market regulations and the corresponding formal and informal institutional framework. Thus, it only provides theoretical guidance on managing the urban-rural water interdependence; and its implementation will be contingent on the actual economic, social and geographical characteristics of the specific water basin. It might therefore be preferable in some cases to dismiss this market model in favour of other types of regulations and policy objectives.

to "rebound effects", i.e. the introduction of the new water-saving technology may trigger behavioural responses on the part of users that cancel out the gains from its adoption.

Rationing policies entail mandatory water-use restrictions in certain areas, under set hydrological circumstances, or for given purposes. Most national or regional authorities in OECD countries have developed priority allocation systems to be applied when water is scarce. While they differ in their application and scope, they essentially consider drinking or domestic use as priority over agricultural (and other) uses (OECD, forthcoming c). In Hungary's Tisza-Koros Valley, drinking water, national security and domestic uses – including medical waters and municipal services – have the highest priority ranking, followed by agricultural (animal husbandry and fish farming), environmental (including environmental protection) and finally industrial and energy production uses (OECD, forthcoming c). Quantitative targets or restriction to superfluous water uses (such as lawn-watering or car-washing) also apply and can be extended to (low-value) agriculture and domestic users in case of increased scarcity. Water restrictions may also force inter-sectoral water reallocations (see the example of Jingmen City, China, in Box 4.3).

The main advantage of rationing policies is their simplicity and ease of implementation. However, they may lack the flexibility required to allocate water efficiently among users and uses. They also require information, notably to define a baseline. During the 2014 drought, California's governor blocked the use of the main canal for agricultural users, imposed mandatory restrictions on the use of water for outdoor activities and called for a 20% reduction in water consumption for all users on a monthly basis. While this rationing

Box 4.3. **Mandatory rural-urban reallocations in the Zhang He Irrigation System, China**

The Zhang He Irrigation System in China's Hubei Province covers a total area of 5 540 square kilometres (km^2). It supplies water to the many rice paddy fields in the region, to the residents of Jingmen City, and to several industrial production activities. Due to the city's rapid urbanisation and industrialisation in the 1990s, the amount of water devoted to residential and manufacturing activities overtook the share allocated to agricultural production. This inter-sectoral redistribution was not the result of free-market interactions between the competing users, but rather of a decision taken at the central level by the regional water operators. The use of water for agricultural production was therefore reduced unilaterally, using a top-down restrictive approach.

The success of this water reallocation policy rests on the fact that the amount of agricultural production in the Zhang He basin did not decrease as a consequence of reduced water allocation; hence, the farmers' livelihoods did not suffer. This is because the farmers adopted new water-saving production methods and technologies, e.g. alternating between wet and dry irrigation methods or building several small reservoirs that captured the run-off and return flows from rice cultivation. In this case, a classical command-and-control policy was successful in driving a welfare-enhancing water transfer between urban and rural users. Moreover, this reallocation stimulus was simple, and its transactions costs low.

Implementing a water-trading model would not have been feasible in this case. In fact, managing an effective mechanism to monitor compliance by the many farmers in the Zhang He basin (who own only a small fraction of the total cultivated land) would have been impossible.

Source : Molden, D. et al. (2010), "Governing to Grow Enough Food without Enough Water – Second Best Solutions Show the Way", *International Journal of Water Resources Development*, Vol. 26(2), pp. 249-263; Bin, D. (2008), "Study on Environmental Implication of water-saving irrigation in Zhanghe Irrigation system", project report submitted to the Food and Agriculture Organization Regional office for Asia and the Pacific, Rome.

policy was voluntary, mandating such a reduction, which would be set against a baseline (past average use) that varies largely among users, would not achieve an efficient allocation under the prevailing water scarcity constraints.

Water conservation measures

Water conservation programmes are demand-side policies, whose final objective is to provide incentives for reducing the amount of water consumed by residential, industrial and agricultural users. One of their greatest advantages is that they reduce dependence on direct abstraction of freshwater supplies and are therefore especially fit for areas where significant political, economic and environmental barriers to procuring new sources of water supplies exist (see Box 4.4 for an illustration in Southern California).

Box 4.4. **Southern California's Water Savings Incentive Program (WSIP)**

Water conservation programmes play a fundamental role in procuring water supplies for many cities in Southern California. The area is under great water stress and is vulnerable to severe droughts. Water conservation has therefore become a tool for curbing water demand in order to hedge the region's demand-supply mismatch. In 1998, the city of San Diego initiated an agreement with compensating Imperial Valley farmers for water conservation measures; this arrangement has become the largest rural-urban water transfer in the United States. In 2011 alone, the water saved by farmers and sent to the city amounted to nearly 100 million cubic metres (m^3). It is projected to reach 237 million m^3 – representing 37 % of San Diego's water supply sources – by 2021.

It has also been estimated that conservation programme activities and water savings initiatives in Southern California have resulted in supplying cities an extra 81 410 m^3 (66 000 acre-feet*) of water a year. One of the water demand management policies that helped achieve these results is WSIP, which targets water savings by businesses, agriculture and large landscapes. The programme is managed by the Metropolitan Water District of Southern California, a consortium of 26 cities and water districts that provides drinking water to nearly 19 million people in parts of Los Angeles, Orange, San Diego, Riverside, San Bernardino and Ventura counties.

The programme provides financial incentives for customised water efficiency projects, paid on the basis of the water amount saved as a result of the projects. The programme pays back up to USD 0.60 per 1 000 gallons saved ever year over the project life, up to a maximum of 10 years, and covers 50% maximum of the incurred costs. Eligible projects include comprehensive changes made to an industrial process that will improve water-use efficiency by reducing water consumption per unit of output, and improvements to existing irrigation systems and plant material changes that improve water-use efficiency for agricultural operations and large landscapes (WSIP Application Package**). These measures diverted water resources from industrial and agricultural users into satisfying residential needs, which are the predominant claim on the area's water resources.

*An acre-foot is defined as the volume of one acre of surface area to a depth of one foot. It is commonly used in the United States to measure large volumes of water (source: Wikipedia).

**www.bewaterwise.com/images/water-savings-incentive-program/Application-Pkg.pdf.

Source: McDonald, R.I., and D. Shemie (2014), *Urban Water Blueprint: Mapping conservation solutions to the global water challenge*, The Nature Conservancy, Washington, DC, http://water.nature.org/waterblueprint/; Richter, B.D. et al. (2013), "Tapped out: how can cities secure their water future?", *Water Policy*, Vol. 15, pp. 335-36; USEPA (2002), "Cases in water conservation: How efficiency programs help water utilities save water and avoid costs", USEPA, Washington, DC; WSIP Application Package, www.mwdh2o.com/, www.bewaterwise.com/.

These programmes may use four types of incentives as leverage for abating water consumption: information incentives, access to conservation technologies, cash transfers and financing (California Urban Water Agencies, 1994). Information incentives include providing free audits or training programmes to make agricultural and residential water consumers aware of the potential monetary savings from adopting water conservation practices. However, in many instances, providing information may not suffice to prompt users to change their behaviour or acquire new water-saving technologies; this is especially true of financially constrained water users. Hence, the government might consider facilitating access to conservation technologies, or lowering the total cost of the water-saving investment through direct cash transfers or other financing mechanisms (e.g. contracting). In the South of France, the Hérault department signed a contract with fourteen municipalities, the chamber of commerce and the chamber of agriculture for them to reduce groundwater use around an aquifer in exchange of shared payments (Barraqué et al., 2010).

The operational decisions taken to implement the various programmes will have a paramount effect on the degree of water-saving achieved. Moreover, since a distinctive feature of these programmes is voluntary participation, correctly marketing the initiative becomes central. Finally, each initiative should be flexible enough to accommodate the changing geographical and economic landscape, and adaptive management should be practised when executing the policy.

Economic instruments: Water pricing and tradable water rights

Economic instruments for distributing water between urban and rural areas require clear price signals providing incentives for individual consumers to use water efficiently. Water pricing underpins the Water Framework Directive 2000/60/EC, which establishes an integrated approach to water management in the European Union through incentive pricing, full-cost recovery and the polluter pays principle. Putting the right price tag on water can be done either through administrative water pricing by a central authority or water-quota trading. Whatever the approach chosen, agricultural water use is substantially under-priced in most OECD countries, as illustrated in Box 4.5 (OECD, 2010).

Water-quota trading can foster an efficient redistribution of water resources across competing users; in fact, standard economic theory predicts that allowing urban and rural consumers trade their water rights in a competitive market would result in welfare-maximising allocation.[2] In other words, water trading leads to gains from trade deriving from the redistribution of water entitlements from low-valued to high-valued users. Moreover, several empirical studies have found that economic instruments such as water pricing or tradable water rights offer a more cost-effective water management strategy than command-and-control policies, thanks to the heterogeneous nature of users' marginal benefits from water consumption (Olmstead and Stavins, 2009).

However, the efficiency gains linked to using economic instruments should also be weighed against the possibility of achieving unfair allocations. The overall equity implications of economic instruments depend on their precise design: whether the water-pricing scheme is linear or block-rate (see Chapter 2 for a more detailed discussion of equity in water tariffs); how water rights are initially allocated; whether water trading may refrain low-income farmers from receiving enough water to sustain their livelihoods in water-stressed regions, etc. They should therefore be considered when determining how to promote the acceptability and long-term stability of the reforms.

Several studies on the potential economic gains from water trading have analysed rural-urban water transfers. They have generally relied on computable general equilibrium models

(CGEs), which take an ex-ante perspective and are thus able to assess the gains from water trading of competing market participants in various contingent scenarios. Even though the studies vary in terms of the specific water basin taken as unit of analysis and the different optimisation techniques used to solve the model, they all reach the same conclusion: in frictionless market settings, unrestricted water-right trading among all the competing users is the policy option yielding the highest possible welfare gains to the population.

Box 4.5. **Full-supply cost recovery for surface water delivered on-farm across OECD countries, 2008**

100% cost recovery of operation and maintenance (O&M) and capital costs: Austria, Denmark, Finland, New Zealand, Sweden and the United Kingdom.

100% cost recovery of O&M, but less than 100% recovery of capital costs: Australia, Canada, France, Japan and the United States.

Less than 100% cost recovery of O&M and capital costs: Greece, Hungary, Ireland, Italy, Mexico, Netherlands, Poland, Portugal, Spain, Switzerland and Turkey.

Less than 100% cost recovery of O&M costs, with capital costs supported: Korea.

Recovery of other costs through water charges or water pricing: opportunity costs, economic and environmental externality costs

- In Australia, some environmental costs have already been recovered, but the opportunity, economic and environmental costs were initially expected to be recovered by 2010.
- France is recovering a share of the environmental costs through water charges.
- The United Kingdom is currently recovering a share of its environmental costs.

Source: OECD (2010), *Sustainable Management of Water Resources in Agriculture*, OECD Publishing, Paris, http://dx.doi.org/10.1787/9789264083578-en.

In the case of the Bow river sub-basin of Southern Alberta, Canada, Ali and Klein (2014) conclude that overall net benefits under unrestricted trading are higher compared to any other possible allocation policy – i.e. seniority rule or people first – as well as under any shortage scenario. Dwyer et al. (2005) use a multi-region comparative static CGE to study the effects of expanding water trade from the irrigation sector to households and other industries in South-eastern Australia; assuming a sudden reduction in available water, they show that the gross regional product would decline by only 0.03 % when full water trading is implemented, but fall by 0.11 % when inter-sectoral trade is not allowed. Other analyses find the same results in many other regions, e.g. Italy and Spain (Pujol et al., 2006), Egypt (Gohar and Ward, 2010), north-central Chile (Hearne and Easter, 1997), the Yellow river basin in China (Heaney et al., 2005) and south Texas (Chang and Griffin, 1992).

Beyond efficiency, water trading also provides the system with the necessary flexibility to adjust drought-induced allocation inefficiencies. Indeed, by allowing trades between urban and rural areas, water markets give access to water resources to users who might have been excluded from them in the event of a drought. In other words, the allocation flexibility embedded in water markets allows them to respond effectively and rapidly to severe supply shocks, such as those caused by droughts (Chong and Sunding, 2006). This welfare-enhancing reallocation is not always possible under a prior tradeless appropriation regime or traditional queuing-allocation system.

Water-trading systems can further enhance their flexibility by introducing dry-year option contracts to secure water access. These adaptive management arrangements allow temporary water transfers that are contingent on water availability and may help mitigate risks related to water scarcity or competition to access the water needed by cities. Such contractual arrangements were used by the Metropolitan Water District of Southern California (MWD), which signed an option contract with the Sacramento Valley irrigation districts in 2003 to secure 180 100 m^3 (146 000 acre-feet) of water; other major agreements involve MWD and Palo Verde Irrigation District, as well as the city of San Diego and Imperial Irrigation District.

From an urban perspective, water markets also allow water allocations to be responsive to the demand stimuli stemming from urban growth. By fostering the emergence of a clear price signal – which will rise in the event that rural areas undergo urbanisation – water markets lead to maximum gains from trades in water transfers. Finally, from the farmers' point of view, flexible water prices create an incentive to adopt new agricultural technologies. For instance, higher water prices stemming from stronger competition for resources may lead to widespread adoption of water-saving irrigation technology. Thus, water prices can also be an effective tool for inducing water conservation (Caswell and Zilberman, 1985). Box 4.6 provides an example of water trading that includes urban and rural users.

Box 4.6. **Water trading in Reus, Spain, and the Siurana-Rudecanyes irrigation district**

The Siurana-Riudecanyes irrigation district is located in the province of Tarragona (Cataluña) and covers 4 000 hectares of land. Since 1904, the stock of water resources in the basin has been allocated through a water market mechanism managed by the Irrigation Subscribers Association of the Riudecanyes Reservoir, an association that includes the city of Reus, the other municipalities and the region's small rural landowners. The association is also responsible for maintenance and repair on the water infrastructure, as well as for financing new water-procurement projects.

The market was instituted with the dual purpose of securing enough water supplies for the city of Reus and preserving the rural areas' agricultural production activities. Water rights were distributed using fixed-price public offerings to members of the Association only, who had to own land in the basin. The collected revenues from these auctions were used to finance dams and other infrastructure. Moreover, the water rights were tradable among the members of the association through both temporal and permanent transfers, and water prices were allowed to fluctuate according to supply conditions.

The market has always been very active, with frequent trades between farmers and municipal water supply companies providing water for domestic and industrial users. The market-allocation mechanism was successful in accommodating both the changing urban-supply water demands and the variations and diversity of the cultivated crops. Hence, the water allocation problem in the Siurana-Riudecanyes basin has been managed in a flexible and economic efficient way through a water-trading mechanism.

Source: website of the association (www.pantaderiudecanyes.cat/); Tarrech R., M. Marino and G. Zwicker (1999), "The Siurana-Riudecanyes irrigation subscribers association and water market system", in Marino, M. and K.E. Kemper (eds.), "Institutional frameworks in successful water markets", *World Bank Technical Paper*, No. 427; and Torregrosa Martí, M.T. (2009), "La gestión del agua en la Marina Baja (Alicante)", *Temas de las cortes valencianas*, University of Alicante, Alicante.

Despite their efficiency advantages, water-trading regimes have not been widely implemented to date. In many OECD countries, water allocation between competing users has remained a bargaining process that is often characterised by a strong political dimension. One reason is that second-best solutions might in some cases provide a more effective (albeit more costly) way to solve the water competition problem, while at the same time addressing issues of fairness and food security.

Another reason is the high cost of entry and institutional complexities involved in establishing water-trading mechanisms; on urban-rural water markets, local rural communities should actively participate in designing the rules and mechanisms regulating water trades. Not only will this ensure a higher probability of their participating in the programme (trades are always voluntary), but the market design will also benefit from local knowledge of the geographical and social characteristics of the water basin.

Groundwater conservation policies

Groundwater represents a major source of domestic water use (about 60% of all drinking water), as well as an increasing share of agriculture use (40% of total irrigation) (OECD, forthcoming b). Its advantages – notably its relative insulation from climatic events and its accessibility to farmers – have led to its intensive use in areas with limited or unstable surface water supplies. This has created tensions with other users, and in some cases observably depleting reserves.

In response to these concerns affecting both urban and rural communities, a number of local management initiatives have been developed around cities, with the objective of finding ways to conserve groundwater supplies. Most do not involve demand-side measures (e.g. restrictions, trading or incentive measures), but rather integrating different water sources and/or supporting practices aiming to replenish groundwater resources. They also involve managing aquifers, and are therefore not always connected directly with managing river basins or larger spatial units. Just as groundwater resource characteristics vary widely, approaches also differ based on uses and institutional constraints (OECD, forthcoming c).

In many cases, cities have offered irrigators the use of treated wastewater in exchange for groundwater for conservation (or banking) purposes. The city of Topeka, Kansas, paid for groundwater rights in exchange for treated municipal wastewater; it fulfilled its conservation objective and reportedly did not use groundwater for seven years (Peck, 2007). Faced with the risk of groundwater depletion, the city of Wichita, also in Kansas, transferred a large volume of surface water into its section of the High Plains Aquifer. Meanwhile, the district of Santa Clara, California, created a conservation district and managed to stop land subsidence that damaged infrastructures; its management plan included monitoring groundwater use among rural and urban users, importing surface water and artificially replenishing the aquifer with treated wastewater (Borchers et al., 2014). Another example is the storage and recovery approach (S&R) in groundwater conservation employed by the city of Tucson, Arizona, which serves the twofold purpose of storing water underground for future use while replenishing the already pumped groundwater (see Box 4.7). This approach also signals a remarkable capacity to adjust to community demands.

Other cases have involved payments or financial transfers. Cities and industrial companies have paid farmers to encourage practices, such as paddy rice flooding, that help recover the aquifer. The city of Ono in Fukui Prefecture, Japan, which received the Japan Water Grand Prize in 2012, was one of the early adopters of groundwater recharge through paddy-field storage in the late 1970s. The city of Kumamoto, Japan, introduced a similar mechanism, acting in conjunction with Sony Corporation and a local foundation to support farmers (Hashimoto, 2013a); the rice was then promoted as environmentally friendly. The Japanese city of Azumino set up a mechanism whereby it paid users to conserve water to benefit urban, industrial and agricultural users (Box 4.8). Other cities have used conservation easements, providing protection to land surrounding the city in exchange for aquifer recharge (see, Lee 2014, for the example of San Antonio, Texas).

Box 4.7. **Tucson's groundwater recharge – S&R approach**

Tucson Water is the city department providing water to most Tucson residents. Its water resources planning and management must meet state groundwater regulations and national quality standards for drinking water. Additionally, it has to comply with the Groundwater Management Act of 1980 (designed to curtail overdraft of aquifers that was occurring in the populated portions of the state, including Tucson) and accompanying regulations that require a demonstration that new residential growth in Tucson will be served by a 100-year assured water supply (AWS).

Besides relying on groundwater, Tucson lacked another source of water supply to comply with the AWS rules. The Central Arizona Project (CAP), a federally-funded canal project that pumped water from the Colorado River from near sea-level to a maximum elevation near Tucson became significant to Tucson Water. Delivery of this renewable water supply provided Tucson Water with a much-needed alternative to groundwater. The region's good-quality groundwater did not require much treatment prior to delivery to customers, but integration of this new water source required building large centralised treatment plant. With the aim of moving away from groundwater overdraft, Tucson delivered treated CAP water – the first real infusion of surface water into the Tucson Water System.

This bold move was fraught with difficulties, related to introducing water with a different chemistry from that of groundwater and travelling in a different direction through old water mains. Issues related to water quality and damages associated with burst pipes, coupled with utility hesitancy to acknowledge the problems, led to a lack of confidence and customer activism to restrict the uses of this new water source. These aspects effectively led to prohibiting the use of the centralised water-treatment plant.

As a consequence, Tucson Water decided to implement an indirect approach to using CAP water. Instead of treating the water in a large storage facility before delivery, it deployed an S&R approach that complied with Arizona State regulations. This approach accomplished more than one goal: *(i)* it avoided the costs of centralised treatment plants; *(ii)* it stored water underground for future use; *(iii)* it replenished groundwater already pumped; and *(iv)* it addressed water management-related objectives. Thus, the S&R approach allowed utilising CAP water, first by storing it underground, mostly in large, shallow spreading basins, where it mixes with groundwater in the aquifer, and then by recovering it for distribution.

Tucson Water's commitment to meet long-term community demands for water, as well as its flexibility and ability to shift from direct delivery to S&R, enabled it to utilise groundwater without expensive large-scale treatment systems and store water for future use. Additionally, Tucson's commitment to conservation recognises water banking as an important strategy for addressing long-term needs for the region.

Source: the case study was developed by Sharon B. Megdal and endorsed by Tucson Water.

Box 4.8. **Groundwater transfer arrangements – Wichita, Kansas, and Azumino, Japan**

The city of Wichita is located above the High Plains Aquifer, a large aquifer that is one of the most intensively used for agriculture in North America. After decades of intensive pumping, especially for irrigation, beginning in the 1940s, the water table declined by up to 13 metres in some locations. In the 1990s, the city started an aquifer S&R programme with the goal of combating the groundwater depletion affecting both urban and regional users. The project is based on recharging the aquifer with treated surface water from the Little Arkansas River during periods of above-normal flows through basin, trench or injection wells. The recharged volume is intended to *(i)* be re-purposed by the city for municipal use; *(ii)* benefit irrigators; and *(iii)* serve as a hydraulic barrier to prevent the intrusion of (natural) saltwater from underground oil-field brine plume. The project entered its second phase in 2009, with the goal of reaching a daily recharge of 378.5 million litres of water.

Japan has long attempted to conserve groundwater not only for the benefit of its largely populated urban area, but also in part to combat land subsidence and salt intrusion in coastal areas. The question of groundwater use recently became problematic in the city of Azumino (Nagano Prefecture) owing to competitive pressure from three sectors: agriculture (through the well-known wasabi production, requiring clean fresh water); industry (through the bottled water sector, which promises pure water sourced in Japan's Northern Alps); and urban residents. A Groundwater Conservation Study Committee was initiated in 2011 to study the problem and propose solutions. Its study, published in August 2012, concluded with the need to increase aquifer recharge. The recharge would be covered by co-operative funds, collected from fees paid by groundwater users depending on the amount of their contribution to the recharge.

Source: Peck, J.C. (2007), "Groundwater management in the High Plains Aquifer in the USA: Legal problems and innovations", in Giordano, M. and K.G. Villholth (eds.) *The Agricultural Groundwater Revolution: Opportunities and Threats to Development*, CABI, Oxford; Sophocleous, M. (2010), "Review: Groundwater management practices, challenges, and innovations in the High Plains Aquifer, USA – Lessons and recommended actions", *Hydrogeology Journal*, Vol. 18, pp. 559-75; Sophocleous, M. (2012), "The evolution of groundwater management paradigms in Kansas and possible new steps towards water sustainability", *Journal of Hydrology*, Vol. 414-415, pp. 550-559; Hashimoto, J. (2013a), "Japanese Municipalities' Efforts to Conserve Groundwater", *Japan for Sustainability Newsletter, No.* 133, September 2013, www.japanfs.org/en/news/archives/news_id034270.html; Hashimoto, J. (2013b), "Municipality's Efforts toward Sustainable Groundwater Use Awarded 15th JWGP Environment Minister's Grand Prize", *Japan for Sustainability*, 8 August 2013, www.japanfs.org/en/news/archives/news_id032985.html.

As schematised in Figure 4.2, the mechanisms typically used involve water transfer, concessions, or payments by cities to rural activities to compensate for the service of ensuring quality groundwater supplies. In some cases where externalities affect both cities and rural areas, contracts involve both types of users – but (as in other cases) agriculture is generally not the basis for such arrangements.

The examples provided in this report cover a limited number of countries and/or regions, partly because cities were able, or were pressured, to act on groundwater management. As happens with surface water, some countries may implement regional or federal schemes that do not give cities any degree of freedom. Groundwater use may also incur limited oversight in many regions due to a lack of investment in monitoring and less pressing groundwater concerns.

Figure 4.2. **Agricultural groundwater management and cities: Water and financial transfers**

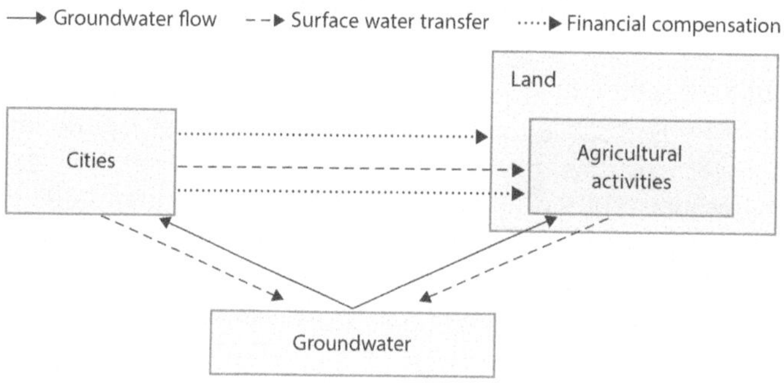

Mitigating flood risks in urban areas

Flood risk mitigation strategies include a set of activities, from risk mapping to emergency plans, land use regulations and hydraulic infrastructures (e.g. dams). In the past decades, the development of hydraulic infrastructures, improved flood risk management plans and better risk preparedness (among other efforts) have allowed many OECD countries to reduce their flood risk vulnerability, despite increased exposure from economic growth and urbanisation. However, recognition of the limitations of such traditional, engineer-based approaches to flood control has led to the gradual emergence of a new paradigm over the last 10-15 years, based on a more integrated and landscape-based approach to flood risk management and a stronger emphasis on ecosystem-friendly strategies.

Both the World Meteorological Organization (WMO) and Global Water Partnership have implemented integrated flood management (IFM) since the early 2000s, building on the pre-existing concept of integrated water resource management (WMO and Associated Programme on Flood Management, 2009). The principle of IFM is to "integrate land and water resource development in a river basin, within the context of Integrated Water Resource Management, with a view to maximising the efficient use of floodplains and to minimize loss of life and property" (WMO, 2009). In the European Union, the trend is to shift priority from "grey" to "green" infrastructures, in close relationship with the concept of IFM, with a particular emphasis on developing natural water resource retention measures (Linnerooth-Bayer et al., 2013) based on soil-management practices. At national levels, typical examples include the Room for River initiative in the Netherlands (see OECD, 2014b).

Rural-urban linkages are likely to be a major dimension of this re-orientation of flood risk management. In certain areas, agricultural land has the potential to play a key role (OECD, 2010; Morris et al., 2010):

1. Agricultural soils can contribute to water retention through the adoption of a set of appropriate farming practices.
2. Agricultural land is also affected by flooding, but can also be used in an integrated landscape perspective for floodwater storage in order to reduce risk in urban or industrial areas and minimise the social cost of flooding.

Forest ecosystems and suburban forests can also play an important role in providing flood control services. They regulate the hydrological cycle by absorbing and storing excess water, decreasing run-off rate and volume, and increasing infiltration and groundwater recharge base flow.

Cities also affect flood risks in their environment: globally, urbanisation has expended sealed surfaces, exacerbating problems with run-offs in case of heavy storms (Swan, 2010). Stormwater management is therefore crucial to mitigating flood risks in urban areas. Box 4.9 analyses the problem and discusses the best practices adopted by cities in Ontario, Canada, echoing similar experiences reviewed in Chapter 3. Finally, the rural environment can also benefit from more sustainable approaches to urban drainage, built into urban planning.

Box 4.9. Stimulating innovative stormwater management in Ontario, Canada

The traditional approach to stormwater management in Ontario is an efficient underground storm sewer network to convey urban run-off as quickly as possible to a nearby water body. In older parts of some cities, such as Toronto's downtown core, stormwater is also conveyed with raw sewage in a combined sewer network to a wastewater treatment plant.

The increase in impervious surfaces, combined with traditional stormwater management, has significantly altered the movement of water in urban areas. The changes in peak flow rates and the total volume, frequency and duration of stormwater have had significant environmental impacts, e.g. bank erosion, and increased the flooding potential.

Additionally, urban stormwater becomes highly polluted as it travels along the urban environment, picking up contaminants including sediment, nutrients, hydrocarbons, heavy metals, road salt, pesticides and animal waste. This polluted water is mainly discharged into an untreated water body; in the case of combined sewers, large storms can result in raw sewage and polluted stormwater bypassing the wastewater treatment facility. Ultimately, the impacts of urbanisation include impaired fish habitat for spawning and rearing, decreased fish health, reproduction and diversity, and unsightly and potentially dangerous algae blooms.

In an attempt to address these issues, Ontario has undergone a series of transformations away from its traditional stormwater management approach focusing on flood control. Stormwater management is now designed for run-off volume, peak flow and quality, taking into account factors such as temperature control, infiltration, water budget and fish habitat. Some practices also address run-off duration and frequency, and are designed with Ontario's four seasons (such as spring snowmelt) in mind.

Ontario promotes the use of a "treatment-train" approach incorporating source (e.g. disconnected downspouts, rain barrels, and rain gardens), conveyance (e.g. swales, exfiltration systems) and end of pipe (ponds, engineered wetlands) control to manage stormwater.

The Ministry of the Environment's Showcasing Water Innovation programme (SWI) has funded 16 stormwater projects to stimulate innovative research and ultimately help municipalities and the province better manage stormwater.

Lake Simcoe Region Conservation Authority was awarded an SWI grant to retrofit existing stormwater ponds to include quality control. It used different retrofitting methods and technologies at three pilot sites. The conservation authority is assessing each of them to determine their effectiveness and applicability for future retrofits. Stormwater ponds have been extensively used throughout Ontario, so the lessons derived from this project will help inform municipalities on how to effectively upgrade their ponds.

Source: WaterTAP (2013), copied from www.watertapontario.com/news/blog/ontario-innovative-stormwater-management-oct2013-/32 (accessed 14 April 2014).

Practical approaches involve providing specific sets of payments for ecosystem services (PES) that reward landowners for providing risk mitigation services through different practices and/or conservation. Box 4.10 illustrates such approaches in several UK towns.

Box 4.10. **Using agriculture fields for flood protection: Examples in the United Kingdom**

The role of agricultural land in flood risk management has been especially underlined in the United Kingdom. Several initiatives and pilot projects have been conducted that include agricultural lands as key players of flood management programmes in the context of the "Making Space for Water" programme. One example is the recent pilot project "Payment for Ecosystem Services (PES) on Flood Regulation" in Hull, whose objectives were to characterise the current state of ecosystem services delivered to urban areas; identify potential improvements of ecosystem service delivery; and design potential payments for ecosystem services. The pilot project led to two proposed PES schemes, including a country park-scale PES that would allow mitigating flood risk in north western Hull by developing "swales, bunds, ponds, replacement of permeable road and car park surfaces and conversion of amenity grassland to semi-natural grasslands and more varied woodlands". Another example is the Beckingham Marshes Washland Creation, which aims to create 94 hectares of floodplain grasslands in order to improve flood risk mitigation for the towns of Gainsborough and Beckingham on the River Trent. While agricultural land areas have contributed to flood risk management in this area since the 1960s, this project also covers natural habitat restoration.

Source: UK Environment Agency. (2010), *Working with Natural Processes to Manage Flood and Coastal Erosion Risk*, UK Environment Agency, Bristol, http://webarchive.nationalarchives.gov.uk/20140328084622/http:/cdn.environment-agency.gov.uk/geho0310bsfi-e-e.pdf; URSUS consulting (2013), *Payment for Ecosystem Services (PES) Pilot on Flood Regulation in Hull*, report for the UK Department of Environment, Food and Rural Affairs, London.

Although these are promising approaches, significant obstacles to their development remain. The first challenge is uncertainty about the linkages between land use, farm practices and flood risks, making it difficult to calibrate the appropriate level of intervention. Hopefully, integrated land-hydrological modelling will provide helpful tools in this area. The second challenge is finding a set of policy tools encouraging farmers to contribute to flood mitigation efficiently. Information about risk levels is a major prerequisite in this regard, as demonstrated for example in the EU Flood Directive 2007/60/EC requiring Member States to map flood risks. Land regulation (such as zoning) will certainly continue to play an important role in this area, but more decentralised and incentive-based policy instruments, e.g. PES, are gaining traction.

Water quality standards and PES

Ensuring minimum quality standards for the water available to urban and agricultural consumers is a matter of great concern, especially to citizens of OECD countries (OECD, 2013a). The demand for water quality is driven mainly by: *(i)* domestic consumers who use water for drinking and fulfilling other basic water needs; *(ii)* water ecosystems and their related environmental, recreational and aesthetic uses (including in cities); and *(iii)* agricultural producers who need specific water quality for their irrigation activities. The needs are clearly very different, leading to misaligned incentives for preserving water quality. Hence, managing the problem in an integrated manner through the urban-rural interface might lead to more effective solutions. Many OECD countries incur significant costs from water pollution; for instance, nitrate and phosphate damage costs an estimated EUR 403-754 million in the Netherlands and EUR 150 million in Spain (OECD 2012b).

The main drivers of the problem are intensive agricultural and farming practices (excessive inputs of nutrients, phosphates and nitrogen fertilisers in agriculture, together with greater nutrient concentration in animal manure), lagging sewage treatment and ineffective management of wastewater resources by domestic and industrial users. On an aggregate level, there have been significant improvements in managing these drivers, as shown by the reduction in average nitrogen balance volumes in OECD countries between 1990 and 2009 (OECD, 2013a).

The main policy objective is to foster the sustainability and environmental protection of water systems without causing significant negative effects on regional economic growth and rural development. When dealing with water quality issues, policy makers will need to reconcile equity and sustainability goals with the economic gains and job creation deriving from the polluting activities.

Policy responses

Possible policies include environmental regulations, economic incentives and information provision – or a mix thereof, particularly: *(i)* stringent point-source pollution standards (uniform command-and-control); *(ii)* payment for water-quality services or performance-based conservation subsidies; and *(iii)* water quality trading. As noted above, standard-setting is usually set at a regional or federal level and rarely involves tailored arrangements between cities and rural areas. (There are exceptions: for instance, New Zealand's regional councils are based on catchment areas, meaning that water-related planning documents are situated at the appropriate scale to manage catchments, regardless of whether they cross rural or urban areas.) This section reviews experience with contractual arrangements that involve paying farmers to improve water quality services, as well as rare water quality trading schemes.

Contractual arrangements and payments for water quality services from municipal water organisations to non-point agricultural pollution sources are policy options that aim to create mutually beneficial changes in land-management uses and practices to achieve both environmental and economic gains. In the case of surface water, they are typically voluntary direct financial arrangements between large cities and rural areas, based on the PES notion. Box 4.11 presents the experience of Munich. Similar but smaller cases can be found in the Netherlands, where water utilities run co-operative agreements with parties of 15 to 20 farmers, including training to reduce input use and stimulation through the "payments by results" principle (i.e. proportional to the reduction in nitrate content of water pumped). These arrangements often involve a small increase in water tariffs to cover farmers' compensations.

Some of the schemes were introduced explicitly to support the application of existing regulations. In the 1990s, the French city of Rennes was confronted with excess nitrate concentration in drinking water. To complement existing tools, notably the EU Nitrate Directive 91/676/EEC, the city developed a programme to reduce nitrate concentration in the water supply by inciting farmers to modify their agricultural practices and purchasing land for reforestation. The programme has allowed reducing nitrate concentration to below 5 milligrammes per litre (mg/l) since 2009, in line with the EU Nitrate Directive requirements. Burik et al. (2011) suggest that the programme was about ten times more cost-efficient than building a water-treatment station.

Box 4.11. **Payments for water quality services – Munich, Germany**

The Mangfall Valley in the Bavarian Alps supplies around 80% of Munich's drinking water (the city has 1.2 million inhabitants). The Valley is predominantly used by farmers and agricultural producers, whose activities were causing slow but significant increases in nitrates (15 mg/l) and pesticides concentrations (0.065 microgrammes per litre) in the city's water resources. To address the issue, in 1991 the municipal water provider Stadtwerke München (SWM) implemented a voluntary payment scheme to encourage local farmers to adopt more sustainable organic farming practices.

After estimating the target area using hydro-geological models, SWM launched a public information campaign targeting 120 farmers (mainly dairy producers). The payments were constructed to cover the expected lost income and investments needed to switch to organic farming; more precisely, farmers received a payment of EUR 280 per hectare per year (ha/year) for the first 6 years after the change and EUR 250/ha/year for the following 12 years.

The programme successfully halved nitrate concentration to 7 mg/l; the price increase for final urban consumers due to the payment scheme (EUR 0.005 per cubic metre [EUR/m^3]) was lower than the avoided cost of water-treatment facilities (EUR 0.23/m^3). Moreover, more than 90% of the farmers adhered to the programme and the Munich area is now considered the largest and more active market for organic farming products in Germany.

One of the key success factors of the programme was the city's strong involvement in purchasing and promoting the organic products from the Mangfall Valley. Not only did the city purchase the organic farming goods to supply its schools and municipal restaurants, it also funded several marketing and advertising campaigns aimed at creating a brand identity for the targeted area's agricultural goods. These measures helped build trust between urban and rural water consumers and – together with a clearly defined set of legal rules regulating organic farming practices in Germany – reduced the contractual and transaction costs of implementing the payment scheme.

Source: Grolleau, G. and L.M.J. McCann (2012), "Designing watershed programs to pay farmers for water quality services: Case studies of Munich and New York City", *Ecological Economics*, Vol. 76(2012), pp. 87-94.

Similar schemes have been implemented to conserve groundwater quality. Some schemes involved land-related transactions: the commune of Saint-Ivy, France, bought former farming land and converted it to woodland to protect the aquifer, partly because landowners in France cannot impose crop choices on tenants (Barraqué et al., 2010). In other cases, the schemes involved designing performance contracts: in Santa Cruz County, California, the Resource Conservation District and the Driscoll's strawberry company piloted a performance-based conservation initiative to monitor and improve groundwater quality, and better manage quantity. The plan includes targets, new monitoring mechanisms and incentive payments for conservation to participating farmers (Levy and Christian-Smith, 2011).

Other recent initiatives have aimed to facilitate dialogue and collaboration among various stakeholders, including rural and urban interests, under the leadership of third-party regional authorities. In New Zealand, the Waikato Regional Council and the Waikato and Waipa River iwi (indigenous peoples) have established a "collaborative stakeholder group" comprising representatives from the dairy and horticultural industries, sheep and beef farmers, the energy sector, local government, public water suppliers and environmental non-governmental organisations. Their purpose is to develop changes to the existing regional water management plan in order to address the adverse effects of discharges to land and water in the Waikato and Waipa river catchments.[3]

Widening the scope of these programmes to entire watersheds upstream of cities whenever and wherever possible could further increase the benefits of contracting. McDonald and Shemie (2014) suggested that focusing on source-watershed conservation strategies can be a cost-effective way to reduce the challenges associated with water sedimentation and nutrient flows. They proposed five strategies – forest protection, reforestation, agricultural best-management practices, riparian restoration and forest-fuel reduction – each of which would be adapted to local circumstances. They found that increasing the rate of adoption of best-management practices on 0.2% of watershed agricultural areas from which water is sourced in the 100 largest cities would result in a 10% sediment reduction in the water supply (constituting the most cost-effective option, as shown in Figure 4.3). The same option is less cost-effective than others when it comes to nutrient reduction (also Figure 4.3). These mechanisms would also reduce costs for water-treatment plant and operations (McDonald and Shemie, 2014).

Figure 4.3. **Cost and effectiveness of water quality gains based on watershed conservation**

10% sediment reduction (upper panel) – 10% nutrient reduction for the 100 largest cities (lower panel)

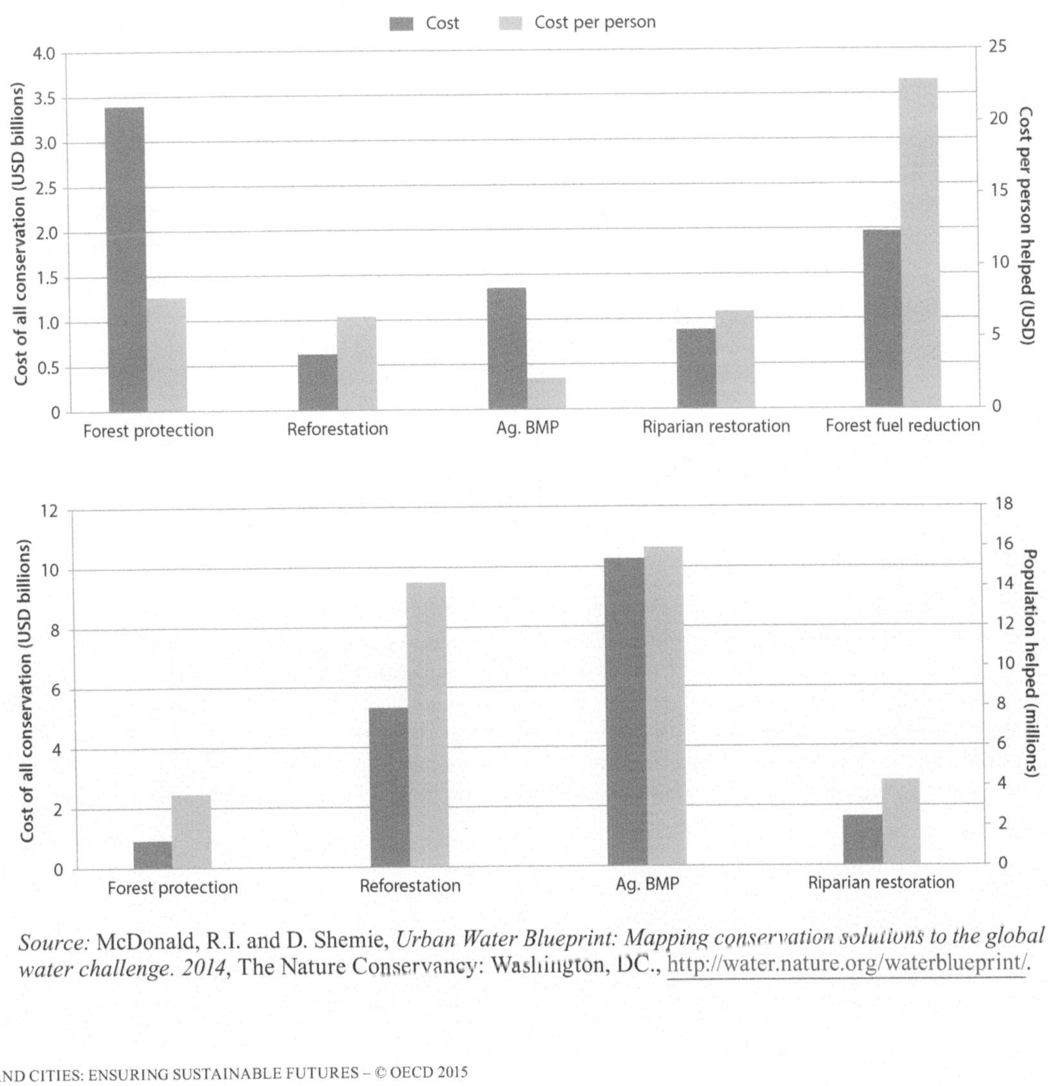

Source: McDonald, R.I. and D. Shemie, *Urban Water Blueprint: Mapping conservation solutions to the global water challenge. 2014*, The Nature Conservancy: Washington, DC., http://water.nature.org/waterblueprint/.

Cities have also applied conservation methods beyond establishing protected areas and other zoning measures. A number of cities have implemented forest protection to ensure water quality (Dudley and Stolton, 2009). In fact, drinking water supplies in 33 of the 105 largest cities – including Tokyo, Los Angeles, Madrid, Vienna or Sydney – depend significantly on protected areas (Dudley and Stolton, 2003). Some cities have deliberately protected their surrounding areas for this purpose, as witness the protected forests around Melbourne, Australia. Others have benefited from protected areas established for other reasons. Even though forest protection may appear to be a minimally cost-efficient solution, it presents considerable (and often non-marketed) co-benefits, especially in terms of additional ecosystem services and biodiversity conservation.

Water quality trading is an application of cap-and-trade pollution control mechanisms that prescribes a maximum level (cap) of total water-polluting diversions allowed for all the subjects participating in the programme. The participants are then allowed to purchase and sell on the market their rights to pollute according to their own necessities, but always within the established cap.

Cap-and-trade economics requires fixing a desired quantity of pollution and allocating quotas to pollution emitters; a common price for water quality emerges from trading these permits. The trading results in assigning the bulk of permits to those participants that can abate their emissions at least cost. If implemented correctly, the system curtails water pollution in the most cost-effective way by ensuring that the marginal costs of pollution reduction are equalised among sources of water pollution. Furthermore, the system produces a great amount of dynamic efficiency by rewarding those who reduce their emissions. In other words, the cap-and-trade mechanism also creates a strong incentive to adopt water-saving and quality-enhancing technologies in order to avoid the higher costs of procurement caused by the resulting emissions' price. Box 4.12 provides an example of a cap-and-trade scheme involving cities at a regional level. Most of these schemes, however, are set at a regional level or between rural water users and do not involve cities (as seen

Box 4.12. **The Pennsylvania nutrient credit trading programme (PANCTP)**

PANCTP targets the flow of nutrients from point to non-point sources from the state of Pennsylvania to the Chesapeake Bay. The Bay's 165 534 km^2 watershed is the largest estuary in the United States and its ecosystems are severely impaired by nutrient pollution. In 2009, Pennsylvania accounted for 44% of the nitrates and 24 % of the phosphates (mainly from agricultural run-offs) that reached the Bay.

Implemented in 2005, PANCTP targets industrial, municipal and agricultural nitrogen and phosphorus polluters. The programme is a partially capped emission-reduction credit trading system that allows uncapped agricultural non-point sources to supply emission credits to capped industrial and municipal sources. Farmers produce emission credits by adopting best-management practices reducing nitrogen and phosphate flows into the Bay. An online market trading platform and a trading clearing house – both managed by a state agency traditionally in charge of financing water infrastructure-development projects – facilitate market-based trading.

While the market started slowly, 259 trades were reported in 2013. The programme's success has not been assessed to date.

Sources: OECD (2012b), *Water Quality in Agriculture, Meeting the Policy Challenge*, OECD Publishing, Paris, http://dx.doi.org/10.1787/9789264168060-en; online market platform (http://pa.nutrientnet.org/); Pennsylvania Department of Environmental Protection, www.depweb.state.pa.us/.

in New Zealand, for instance). Moreover, water quality trading programmes conducted in several countries around the world have had mixed outcomes to date (Shortle, 2012). Different barriers to their development have been identified (i.e. the need for transparent and clear trading rules; significant differences in abatement costs; the need for dedicated institutions to manage the programmes), and implementation is especially difficult in the case of non-point source pollution. Water quality trading remains a complex arrangement, which requires *(i)* the existence of strong management capacity at water basin scale; and *(ii)* the capacity to monitor the operation of the trading mechanisms; the concrete modalities of water quality trading are beyond the scope of this report (see OECD, 2012b, for more information on water quality trading).

Implementation issues

This chapter's review of policy options has highlighted a possible mismatch – at least from an efficiency perspective – between proposed first-best solutions and implemented approaches. For instance, as noted in the section on rural and urban water competition, water-trading systems, despite their recognised economic efficiency, are mainly used intrasectorally (OECD, 2010) and rarely driven or used by cities. This discrepancy may originate from multiple factors, mostly related to ease of implementation: the cost of entry and the institutional complexity of the approaches may make them unfit for smaller-scale operations or difficult to establish among users. Moreover, policy makers often decide to favour other objectives – i.e. equity or sustainability – over economic efficiency. Other global, national or local political objectives may also affect implementation: trade liberalisation, food or energy security, land grabbing, poverty alleviation or the strong lobby groups might tilt the scales towards a particular policy choice.

Two underlying questions best synthesise these issues in the context of rural-urban linkages: what is the appropriate scale for urban-rural co-operation over water management, and what implementation challenges do cities and rural actors face when considering optimal co-operation mechanisms?

Scale of urban water policies and implication for rural-urban co-operation[4]

As discussed in other parts of this chapter, it is worth asking whether and when cities and rural actors should be involved in managing water beyond the existing policies set at a higher level. The answer clearly depends on the scope of the quality, scarcity or flooding constraints they face. Cities centrally located in a water basin are more inclined to see the benefits of rural-urban linkages, even when implemented at a regional scale. Cities with a limited surface water supply that depend on groundwater or face salinisation and/or land subsidence risks have a direct interest in linking with other users of underlying aquifers. By contrast, cities with no flood risks and high quality, steady water allocation have very limited incentives to act. The adequacy and efficacy of prevailing policies also play a role: whether cities become involved will depend on the costs of co-operation arrangements, which themselves depend on economies of scale. Rural-urban co-operation can entail excessive transaction costs and administrative burdens. Co-ordination across levels of government requires time, resources and capacities, which may not be in place when the co-operation process begins.

At the same time, a city will only consider taking action if it has the freedom or authority to do so. Pre-existing conditions may allow them to manage water challenges at a specific level or following an established operational framework that grants limited freedom to cities, e.g. the EU Water Framework Directive 2000/60/EC requires all water stakeholders at the river basin level to co-operate on reaching specific objectives set in their management plans. Other institutional contexts may allow or require cities to manage issues at their own level, with the regional or national authority possibly playing an external role to ensure that rural-urban co-operation linkages are balanced and functioning. The issue of decentralisation is therefore critical to understanding whether and how cities should link with rural areas on issues related to water management. It should be noted that such co-operation mechanisms can be introduced in centralised or federal states as shown by the various examples in the chapter (e.g. France, Japan, Germany and the United States), the question is more whether the institutional system allows (or encourages) cities or regions to take action on water management.

The type of instruments will also depend on the operational scale. First-best instruments may involve entry costs or cause frictions that make them much less affordable than more ad hoc contractual solutions. Taking into account pre-existing regulatory frameworks that address the core constraints of water quantity and quality, the marginal benefit of action may not cover its significant marginal cost. In this setting, ad hoc small-scale contractual arrangements – which have the advantage of addressing a particular city's specific constraints, and involve a limited number of actors – may have much more appeal than well-defined but complex or challenging institutional responses.

Implementing rural-urban policies

The second issue, related to actual policy implementation, revolves around the willingness and means of rural actors and cities to establish linkages. As noted in a number of examples, cities are generally the drivers of urban-rural water linkages.

Rural-urban partnerships can help cities and their hinterlands integrate water management. As defined in Box 4.13, they consist of cross-sectoral and holistic sets of initiatives (e.g. within a wider package of environmental-policy initiatives) or focus on single objectives/projects (i.e. managing water resources). For instance, a rural-urban partnership can aim to manage the production and distribution of benefits associated with ecosystem services. In Forli-Cesena, Italy, water resources are managed through a partnership among all urban and rural municipalities, as well as the chambers of commerce of three different provinces, which are also included in the co-operation process. The municipalities where the water sources are located share in the revenues from water provision, as well as benefit from investments in natural and cultural-heritage preservation and initiatives to develop area tourism. The other municipalities benefit from the availability of clean water and proximity to high-value landscape and amenities (OECD, 2013b). Box 4.14 provides an example of a rural-urban partnership.

In most examples involving agriculture, cities need to provide sufficient incentives for farmers to participate. Such incentives may result from external institutional constraints (e.g. regional or national requirements to act) or external physical constraints (climatic events, saline-water intrusion, and rapid depletion of the water stock) that may be sufficient to trigger co-operative action for the sake of protecting a common interest. In the absence of shared constraints, however, cities will need to be pro-active in defining the proposed mechanisms. Once again, payments for water-related services appear as the most obvious mechanism to do so, possibly focusing on changing agriculture or water-using practices. Trading mechanisms – which may involve other actors – also set a price that is bound to facilitate exchanges. Other reported mechanisms involve non-pecuniary benefits for farmers, including water transfers or groundwater conservation schemes that are mutually beneficial and enhance water security for both parties.

Box 4.13. Rural-urban partnership

Rural-urban partnership is defined as the co-operation mechanism that manages linkages to reach common goals and enhance urban-rural relationships. Rural-urban partnerships are a possible response to gaps in existing levels of governance of urban-rural relationships. They can help deliver services and public goods, as well as develop public goods and improve administration, focusing on spaces of functional territorial integration.

The concept of rural-urban partnership has distinct features, involving:

- an awareness of the interdependence of rural and urban areas in a given space (functional region)
- a membership mix that includes the relevant rural and urban representatives
- a framework for action or objectives that represents mutual interests (urban and rural)
- initiatives aimed at yielding collective benefits for urban and rural partners
- an organisational form that is fit-for-purpose to help realise the partnership's objectives.

One of the characteristics distinguishing rural-urban partnership from other types of territorial co-operation is the fact that both urban and rural areas must be directly involved in the process. This implies including urban and rural stakeholders, such as public authorities (e.g. urban and rural municipalities) and/or private agents (e.g. firms and civil society). Common to all rural-urban partnerships is a common set of objectives, to be managed jointly, in a space where urban and rural dimensions are physically and/or functionally integrated.

Because they are driven by linkages between urban and rural areas, rural-urban partnerships reflect the existence of complementarities, which in turn allow territories to join efforts and resources to reach common objectives that cannot be achieved (or at least not as effectively) in isolation.

Rural-urban partnerships can be **explicit** (if the rural-urban dimension is very clear, and the relations between the urban and rural stakeholders, and their interests, are taken into account in the partnership's membership mix, work and strategic objective) or **implicit** (if the partnership aims to improve co-operation through a common local development objective, strategy or project, but still involves the urban and rural authorities).

Three different designs can shape co-operation:

- **Formal and institutionalised**: the actors involved have a formal commitment to reach out across their respective responsibilities and interests and co-operate on certain issues, integrating activities that were formerly carried out individually.
- **Formal but not institutionalised**: the actors enter into a hybrid partnership with both formal and informal components. The partnership has all the characteristics of the first group, except that it is not institutionalised, and is looser and less structured. It has no independent structure, with staff or allocated resources.
- **Informal**: the members decide to join together in loose networks that permit mutual consultation and co-ordination. No particular body is established, rules for co-operation are not well developed, and competences are limited.

Source: OECD (2013b), *Rural-Urban Partnership: An Integrated Approach to Economic Development*, OECD Publishing, Paris, http://dx.doi.org/10.1787/9789264204812-en.

Box 4.14. The New York City watershed programme: An example of a successful rural-urban partnership

New York City has one of the few sources of natural, unfiltered water in the United States. It is the largest water system in the country operating under an approved filtration-avoidance waiver. A full 90% of the city's water comes from the 4 140 km^2 (1 600 square-mile) Catskill/Delaware Watershed, consisting primarily of rural area – farms, forests and small towns – with a growing number of suburban developments and vacation homes.

A memorandum of understanding (MOU) was signed in 1997 by New York City, communities of Catskill/Delaware watershed, the US Environmental Protection Agency (USEPA), New York State and environmental organisations with the dual goals of "protecting water quality for the generations to come and preserving the economic vitality of watershed communities". The agreement established an institutional framework to implement a range of commonly agreed protection programmes Since its inception, the Department of Environmental Protection (DEP) and its partners have focused on several key watershed-protection initiatives: implementing the Watershed Agricultural Program; acquiring watershed lands; enforcing updated Watershed Rules and Regulations; and initiating and expanding environmental and economic partnership programmes targeting specific sources of pollution in the watershed.

The instituted Watershed Agricultural Council assisted agricultural and forestry communities with adopting management techniques to protect water quality and enhance economic viability. It also supported conservation easements that provided landowners with annual payments in exchange for maintaining the land in a natural state.

One of the key factors supporting the adoption of the watershed agreement was the reciprocal recognition of the legitimate interest of New York City residents for the conservation of water quality and of the ability of Catskills farmers to implement a satisfying environmental programme satisfying this objective. The innovative combination of watershed programmes and partnerships contributed to creating and maintaining the reputation of New York City's drinking water as one of the finest supplied in the United States.

Source: adapted from New York Department of Environmental Protection (2011), *Long-term water protection plan 2011*, New York, www.nyc.gov/html/dep/pdf/watershed_protection/2011_long_term_plan.pdf; and Appleton, A. (2002), "How New York City Used an Ecosystems Services Strategy Carried Out Through an Urban-Rural Partnership to Preserve the Pristine Quality of its Drinking Water and Save Billions of Dollars", paper for Forest Trends conference in Tokyo, New York, http://ecosystemmarketplace.com/documents/cms_documents/NYC_H2O_Ecosystem_Services.pdf.

Beyond providing economic or tangible benefits to engage participants, lasting management schemes may need to rely on cities' capacities to build trust with their rural counterparts. Successful rural-urban partnerships require mutual trust and a clear understanding of the long-term benefits of the interaction. Differences between rural and urban areas in terms of capacity, economic and political power – from the lack of information, evidence and data, as well as the lack of capacity in rural areas – can complicate the relationship. The institutional framework (i.e. regulatory and political barriers) can sometimes constrain rural-urban partnerships, and the absence of proper mechanisms or incentives can undermine co-operation even when there is interest on both sides. The involvement of higher-level authorities through national regulations or guiding principles (e.g. the polluter pays principle) may help overcome barriers to co-operation; so does co-operating on multiple issues (beyond water), involving rural and urban actors as a single undertaking.[5]

Several other general conditions need to be fulfilled for a successful partnership. These include information-sharing before, during and after the partnership (Box 4.15). Early-stage stakeholder engagement and participation are prerequisites for building long-lasting trust. Involving third parties as observers, participants or arbitrators, and ensuring transparency (by setting measurable objectives, monitoring efforts and reporting results) are also critical. More generally, rural-urban arrangements require the same good governance principles that apply to other levels of water management in order to thrive.

Box 4.15. **Conditions for successful rural-urban partnerships**

A successful rural-urban partnership (including for water) relies on five pillars, in which local government plays a crucial role:

- **Promote a better understanding of socio-economic conditions in urban and rural areas**: local government can use the rural-urban governance framework to promote water quality, reduced water consumption and flood protection.
- **Address territorial challenges with an approach based on *functional* linkages between urban and rural areas**: the local government can identify rural-urban areas' strengths and weaknesses in managing water, as well as interdependencies.
- **Encourage integration of urban and rural policies by working towards a common agenda**: local government can encourage different government levels to participate in rural-urban partnerships to achieve better policy integration and align interests inside and outside of the water box.
- **Promote an enabling environment for rural-urban partnerships**: local government can develop trust and a shared vision of the territory by promoting pilot projects on easy "win-win" issues, education initiatives and dialogue facilitators, and setting balanced "rules of the game" (i.e. promoting a fair partitioning of voting rights within the partnership).
- **Clarify the partnership objectives and related measures to improve learning and facilitate the participation of key urban and rural actors**: local government can facilitate the exchange of good practices and knowledge acquired though the rural-urban partnership, promoting evaluation of the initiatives and information-sharing.

Source: adapted from OECD (2013b), *Rural-Urban Partnership: An Integrated Approach to Economic Development*, OECD Publishing, Paris, http://dx.doi.org/10.1787/9789264204812-en.

Notes

1. While this chapter focuses on agriculture because of its primordial importance in water uses, it should be acknowledged that rural areas have mixed economies that are in some cases similar to those in urban areas.
2. Standard economic theory assumes that environmental and social externalities are (or can be) fully internalised, an assumption that seldom materialises.
3. For more information, see www.waikatoregion.govt.nz/healthyrivers.
4. Chapter 5 addresses more generally the challenges associated with multi-level water governance in cities. This section only delves into issues related implementing policy instruments to tackle the three water-quantity and quality constraints.
5. Game theory has shown that enlarging the bundle of negotiated items can help players find a solution.

References

Ali, M. K. and K.K. Klein (2014), "Implications of current and alternative water allocation policies in the Bow River Sub Basin of Southern Alberta", *Agricultural Water Management*, Vol. 133, pp. 1-11.

Appleton, A. (2002), "How New York City used an ecosystem services strategy carried out through an urban-rural partnership to preserve the pristine quality of its drinking water and serve billions of dollars and what lessons it teaches about using ecosystem services", presentation given at Katoomba Conference, Tokyo.

Barraqué, B. et al. (2010), *Groundwater in the Southern Member States of the European Union: an assessment of current knowledge and future prospects. Country report for France*, European Academies of Sciences Advisory Council, Brussels.

Bin, D. (2008), "Study on Environmental Implication of water-saving irrigation in Zhanghe Irrigation system", project report submitted to the Food and Agriculture Organization Regional office for Asia and the Pacific, Rome.

Borchers, J. W. et al. (2014), *Land Subsidence from Groundwater Use in California. Summary Report*, California Water Foundation and Ludhorf and Scalmarnini Consulting, Sacramento.

Burik, B., J. Gault and F. Bertoye (2011), *Payment for Environmental Services: First Global Inventory of Schemes Provisioning Water for Cities*, United Nations Food and Agriculture Organization, Rome.

California Urban Water Agencies (1994), *A Guide to: customer incentives for water conservation*, United States Environmental Protection Agency, Office of Water, Washington, DC.

Caswell, M., Zilberman D. (1985), "The choice of irrigation technologies in California", *American Journal of Agricultural Economics*, Vol. 67(2), pp. 224-234.

Chang, C. and R.C. Griffin (1992), "Water marketing as a reallocative institution in Texas", *Water Resources Research*, Vol. 28 (3), pp. 879-890.

Chong, H. and D. Sunding (2006), "Water Markets and Trading", *Annual Review of Environment and Resources*, Vol. 31, pp. 239-264.

Civita, M.V., A. Massarutto and G. Seminara (2010), *Groundwater in the Southern Member States of the European Union : an assessment of current knowledge and future prospects. Country report for Ita*ly, European Academies of Sciences Advisory Council (EASAC), Brussels.

Colby (1990) "Enhancing instream flow benefits in an era of water marketing", *Water Resources Research*, Vol. 26, pp. 1113-1120.

Dodds, W. K. et al. (2009), "Eutrophication of U.S. freshwaters: analysis of potential economic damages", *Environmental Science Technology*, Vol. 43(1), pp. 12-19.

Dudley, N. and S. Stolton (2008), "Drinking Water and Protected Areas" in United Nations Convention on Biological Diversity (CBD) (ed.), "Protected Areas in Today's World: Their Values and Benefits for the Welfare of the Planet", *CBD Technical Paper*, No. 36, CBD, Montreal, www.cbd.int/doc/publications/cbd-ts-36-en.pdf.

Dudley, N. and S. Stolton (2003), *Running Pure*, World Wildlife Fund, Gland, and World Bank, Washington, DC.

Dwyer, G. et al. (2005), "Integrating rural and urban water markets in south east Australia", paper presented to the OECD Workshop on Agriculture and Water: Sustainability, Markets and Policies Adelaide, 14-18 November 2005, Productivity Commission, Canberra, www.pc.gov.au/research/completed/water-markets/watermarkets.pdf.

Gohar, A.A. and F. A. Ward (2010), "Gains from expanded irrigation water trading in Egypt: an integrated basin approach", *Ecological Economics*, Vol. 69(2012), pp. 2535-2548.

Grolleau G., and L.M.J. McCann (2012), "Designing watershed programs to pay farmers for water quality services: Case studies of Munich and New York City", *Ecological Economics*, Vol. 76 (2012), pp. 87-94.

Groot, N. (2013), "Norm Groot: Projects address seawater intrusion", *The Californian*, 16 December 2013, www.thecalifornian.com/story/news/2013/12/16/norm-groot-projects-address-seawater-intrusion/4036377/.

Hashimoto, J. (2013a), "Japanese Municipalities' Efforts to Conserve Groundwater", *Japan for Sustainability Newsletter 133*, September 2013, www.japanfs.org/en/news/archives/news_id034270.html.

Hashimoto, J. (2013b), "Municipality's Efforts toward Sustainable Groundwater Use Awarded 15th JWGP Environment Minister's Grand Prize", *Japan for Sustainability*, 8 August2013, www.japanfs.org/en/news/archives/news_id032985.html.

Heaney, A. et al. (2005), "Water reallocation in northern China: towards more formal markets for water", in Willett, I.R. and Zhanyi Gao (eds.), *Agricultural water management in China, Proceedings of a workshop held in Beijing, China, 14 September 2005*, ACIAR Proceedings No. 123, Canberra.

Hearne, R. R. and K.W. Easter (1997), "The economic and financial gains from water markets in Chile", *Agricultural Economics*, Vol. 15, pp. 187-99.

Lee, L. (2014), "Rural conservation ensures cities' water supplies, benefits landowners", tx H2O, Summer 2014, Texas Water Resources Institute, Texas A&M University, http://twri.tamu.edu/publications/txh2o/summer-2014/protect-our-land-protect-our-water/.

Levy, M. and J. Christian-Smith (2011), *Groundwater Management in the Pajaro Valley*, Pacific Institute, Oakland.

Linnerooth-Bayer, J. et al. (2013), "Climate change mainstreaming in agriculture: Natural water retention measures for flood and drought risk management," *Policy Update No. 7*, FP7 RESPONSES Project – European Responses to Climate Change, International Institute for Applied Systems Analysis (IIASA), Vienna, www.responsesproject.eu/pdf/D4.4%20PU%207%20project%20report.pdf.

McDonald, R.I. and D. Shemie (2014), *Urban Water Blueprint: Mapping conservation solutions to the global water challenge*, The Nature Conservancy, Washington, DC, http://water.nature.org/waterblueprint/.

Molden, D. et al. (2010), "Governing to Grow Enough Food without Enough Water – Second Best Solutions Show the Way", *International Journal of Water Resources Development*, Vol. 26(2), pp. 249-263.

Morris, J., T. Hess and H. Posthumus (2010), "Agriculture's Role in Flood Adaptation and Mitigation: Policy Issues and Approaches", in *Sustainable Management of Water Resources in Agriculture*, OECD Publishing, Paris, http://dx.doi.org/10.1787/9789264083578-en.

OECD (forthcoming a), *Water Governance in Cities*, OECD Publishing, Paris.

OECD (forthcoming b), *Groundwater Use in Agriculture in OECD countries: From Economic Challenges to Policy Solutions*, OECD Publishing, Paris.

OECD (forthcoming c), *Water Resources Allocation: Sharing Risks and Opportunities,* OECD Publishing, Paris, http://dx.doi.org/10.1787/9789264229631-en.

OECD (2014a), "Global irrigation water demand projection to 2050: An analysis of convergences and divergences", paper presented at the Working Party on Water, Biodiversity and Ecosystems in April 2012, OECD, Paris, www.oecd.org/officialdocuments/publicdisplaydocumentpdf/?cote=ENV/EPOC/WPBWE(2012)2/FINAL&docLanguage=En.

OECD (2014b), *Water Governance in the Netherlands: Fit for the Future?*, OECD Publishing, Paris, http://dx.doi.org/10.1787/9789264102637-en.

OECD (2013a), *OECD Compendium of Agri-environmental Indicators*, OECD Publishing, Paris, http://dx.doi.org/10.1787/9789264186217-en.

OECD (2013b) Rural-Urban Partnership: An Integrated Approach to Economic Development, OECD Publishing, Paris, http://dx.doi.org/10.1787/9789264204812-en.

OECD (2012a), *Environmental Outlook to 2050: The Consequences of Inaction*, OECD Publishing, Paris. http://dx.doi.org/10.1787/9789264122246-en.

OECD (2012b), *Water Quality in Agriculture, Meeting the Policy Challenge*, OECD Publishing, Paris, http://dx.doi.org/10.1787/9789264168060-en.

OECD (2010), *Sustainable Management of Water Resources in Agriculture,* OECD Publishing, Paris, http://dx.doi.org/10.1787/9789264083578-en.

Olmstead, S. (2010), "The Economic of managing scarce water resources", *Review of Environmental Economics and Policy*, Vol. 4(3), pp. 179-198.

Olmstead, S. and R. Stavins (2009), "Comparing price and non-price approaches to urban water conservation", *Water Resource Research*; doi:10.1029/2008WR007227.

Peck, J.C. (2007), "Groundwater management in the High Plains Aquifer in the USA: Legal problems and innovations", in Giordano, M. and K.G. Villholth (eds.) *The Agricultural Groundwater Revolution: Opportunities and Threats to Development*, CABI, Oxford.

Pujol J., M. Raggi and D. Viaggi (2006), "The potential impacts of markets for irrigation water in Italy and Spain: a comparison of two study areas", *The Australian Journal of Agricultural and Resource economics*, Vol. 50, pp. 361-380.

Richter, B.D. et al. (2013) "Tapped out: how can cities secure their water future?", *Water Policy*, Vol. 15, pp. 335-36.

Shortle, J. (2012), *Water Quality Trading in Agriculture: Meeting the Policy Challenge*, OECD Publishing, Paris, http://dx.doi.org/10.1787/9789264168060-en.

Sophocleous, M. (2010), "Review: Groundwater management practices, challenges, and innovations in the High Plains Aquifer, USA – Lessons and recommended actions", *Hydrogeology Journal*, Vol. 18, pp. 559-75.

Sophocleous, M. (2012), "The evolution of groundwater management paradigms in Kansas and possible new steps towards water sustainability", *Journal of Hydrology*, Vol. 414-415, pp. 550-559.

Swan A. (2010), How increased urbanisation has induced flooding problems in the UK/ a lesson for African cities?, *Physics and Chemistry of the Earth*, Vol. 35, pp. 643-647.

Tarrech R., M. Marino and G. Zwicker (1999), "The Siurana-Riudecanyes irrigation subscribers association and water market system", in Marino, M. and K.E. Kemper (eds.), "Institutional frameworks in successful water markets", *World Bank Technical Paper*, No. 427.

Torregrosa Martí, M. T. (2009), "La gestión del agua en la Marina Baja (Alicante)", *Temas de las cortes valencianas*, University of Alicante, Alicante.

UK Environment Agency (2010), "Working with Natural Processes to Manage Flood and Coastal Erosion Risk", UK Environment Agency, Bristol, http://webarchive.nationalarchives.gov.uk/20140328084622/http:/cdn.environment-agency.gov.uk/geho0310bsfi-e-e.pdf.

URSUS consulting (2013), *Payment for Ecosystem Services (PES) Pilot on Flood Regulation in Hull*, report for the UK Department of Environment, Food and Rural Affairs, London.

USEPA (2002), "Cases in water conservation: How efficiency programs help water utilities save water and avoid costs", United States Environmental Protection Agency, Washington, DC.

USEPA (2014) "Urban waters", webpage, www2.epa.gov/urbanwaters (accessed 24 September 2014), United States Environmental Protection Agency, Washington, DC.

Water TAP (2013), "Lake Simcoe Project: Stimulating Innovative Stormwater Management", www.watertapontario.com/news/blog/ontario-innovative-stormwater-management-oct2013-/32.

World Meteorological Organization and Associated Programme on Flood Management (2009), "Integrated Flood Management Concept Paper", *WMO* no. 1047.

Chapter 5

Governance for urban water management

This chapter explores governance arrangements to manage water in cities in OECD countries. Building on previous OECD work on water governance and on two OECD surveys carried out in 2013-14, it identifies governance gaps to urban water management, trends in institutional organisation and policy tools to address them.

The chapter explores in particular three main governance arrangements for responding to the challenges of a complex and fragmented sector: metropolitan governance, regulatory bodies for water supply and sanitation services, and stakeholder engagement.

Key messages

Cities face seven major governance bottlenecks to effective water policy design and implementation:

- *Administrative gap*: the ecological dimension of water cuts across spatial scales, but institutional, functional and hydrological logics affect its governance in cities.
- *Information gap*: information asymmetries and difficulty in collecting and sharing data can affect the decision-making process.
- *Policy gap*: several policy areas influence water governance in cities; policy coherence is often overlooked.
- *Capacity gap*: local actors' limited scientific, technical and financial capacities make it difficult to implement water policies and strategies properly.
- *Funding gap:* unstable or insufficient revenues undermine the effective implementation of water responsibilities at the sub-national level.
- *Objective gap*: conflicting objectives across water uses (agriculture, energy, etc.) and stakeholders can compromise long-term targets for integrated urban water policy.
- *Accountability gap:* difficulties in ensuring transparent practices across the different constituencies affect engagement, deliberation and decision-making.

A range of governance arrangements can help cities and national governments overcome these obstacles. The chapter explores three such arrangements: metropolitan governance, dedicated regulatory bodies and stakeholder engagement.

Urban water management will increasingly combine multiple scales, from urban-rural co-operation to small-scale water services. As a mechanism to pool resources and capacity across municipalities within the metropolitan area, metropolitan governance can help handle interdependencies across authorities and reduce fragmentation to manage water resources and water services more efficiently. In a context rife with financial constraints, this form of governance arrangement gains traction. Central governments have a role to play in incentivising cities to explore it more systematically among the range of options for co-ordinating cross-scale water management.

The establishment of dedicated regulatory bodies for water and wastewater services is a consistent trend in a number of OECD countries. Where established, they are seen as an institutional arrangement that can address several of the governance gaps identified above, promoting transparency; policy coherence and co-ordination; continuity, predictability and credibility of decision-making; and accountability to users. The OECD survey on the governance of water regulators (OECD, forthcoming a) indicates that regulators are generally well-equipped to achieve their objectives, and benefit from their relatively recent establishment and the experience of their peers in other utility sectors.

Stakeholder engagement is increasingly acknowledged as a means to secure willingness to pay for water services; raise awareness of current and future water challenges; ensure the accountability of city managers and service providers to end users and citizens; manage water allocation conflicts; guarantee the political acceptability of different ownership models; and set convergent objectives across policy areas. This chapter proposes several tools and practices to foster stakeholder engagement with urban water management in OECD countries.

Preliminary remark

The chapter builds on the results and findings of two OECD surveys carried out in 2013-14 to document governance trends and arrangements for managing urban water:

- A Survey on Water Governance in cities was conducted across 50+ cities above 500 000 inhabitants, covering urban water governance drivers and challenges, and policy responses to fragmentation across people, places and policies. The following analyses are based on the responses provided by the first 30 respondents.
- A Survey of Water Regulators was carried out between September 2013 and September 2014; it was completed by 34 bodies established to regulate the provision of urban drinking water and wastewater services (list provided in Annex 5.A2). The questions were based on the OECD Best Practice Principles for the Good Governance of Regulators (OECD, 2014). A dedicated report based on the survey describes the features of regulatory bodies, including their functions and powers, institutional setting and internal organisation (OECD, forthcoming a).

Governance challenges to urban water management

Water management cuts across multiple scales, levels of government and policy areas. To manage water within a whole-of-government approach, cities need to pay particular attention to seven categories of governance gaps in order to achieve effective water policy design and implementation (Figure 5.1).

Figure 5.1. **OECD multi-level governance framework: Mind the gaps, bridge the gaps**

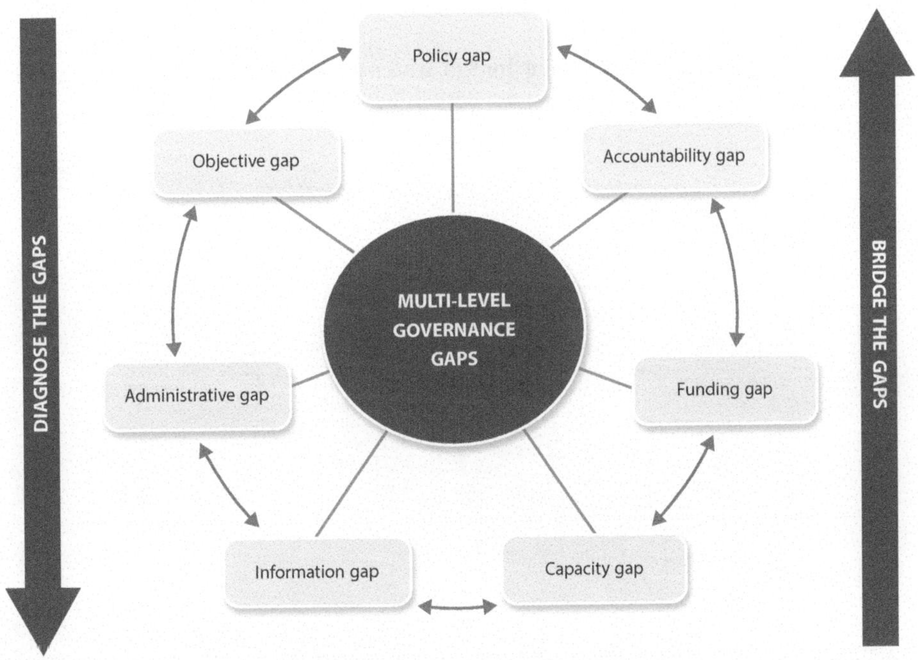

Source: OECD (2011), *Water Governance in OECD countries. A multi-level approach*, OECD Publishing, Paris, http://dx.doi.org/10.1787/9789264119284-en.

Multi-level governance gaps should be considered systemically, as they are deeply interrelated and can exacerbate one another. For example, any city facing a sectoral fragmentation of water roles and responsibilities across public actors (policy gap) may also suffer from conflicting goals (objective gap). Because of silo approaches, policy makers may not willingly share information (information gap). This, in turn, undermines capacity-building at the sub-national level (capacity gap) because local actors, users and private actors have to multiply their efforts to identify the right interlocutor in the central administration. Diagnosing the gaps is a primary step to overcoming obstacles and promoting more effective water policy and management.

The question of scale

An administrative gap occurs when there is a geographical mismatch between hydrological and administrative boundaries. This mismatch can have consequences on the competition over water uses and the effectiveness of service delivery and investment. In administrative terms, **cities** are defined and established by legal actions that serve administrative and governmental functions. While the administrative units are frequently those for which policy is implemented, they are often arbitrary and reflect ancient patterns.[1] Cities can be part of **metropolitan areas**, encompassing different municipalities that are socio-economically closely connected to the central city (or cities) or included in functional urban areas, representing the integration between (large) cities and rural surroundings (Figure 5.2) (Chapter 1).

Functional geographies depend on the function in question. In the case of water resource management, appraising the metropolitan and hydrological logistics is key to addressing linkages between urban areas (where most people live) and the surrounding environment (i.e. rural area and watersheds) that sustains them. On the one hand, functional urban areas

Figure 5.2. **Scalar dimensions of water governance in cities**

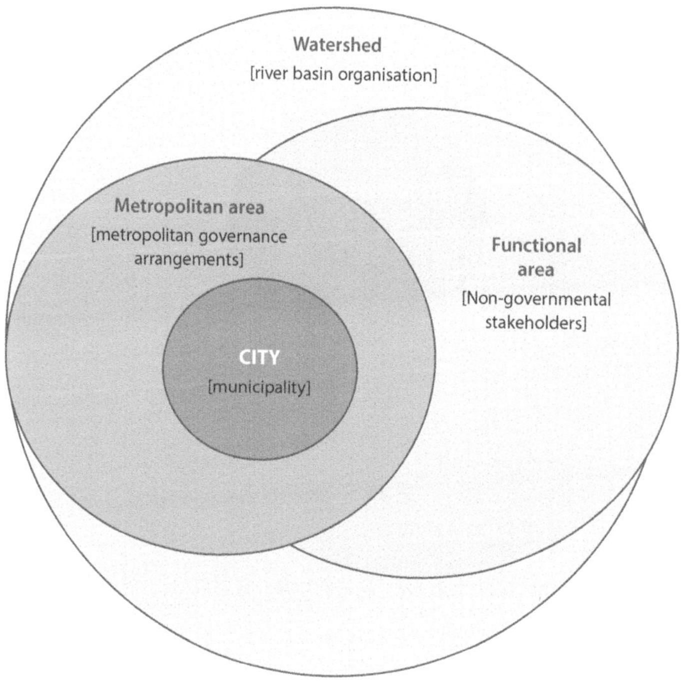

can be used to map centres of urban water demand (and appraise water-related investment needs beyond traditional city boundaries). On the other hand, catchments and basins are relevant to understanding the hydrology (see Chapter 1).

Weak articulation between institutional, functional and hydrological logics affects urban water management. A majority of respondents to the OECD survey (57%) identified the lack of relevant scale for investment as the most critical administrative obstacle to effective urban water governance. Chapter 2 has noted attempts to find the right scale for urban water management by amalgamating local authorities. Inter-municipal co-operation is also widely used across OECD countries to reach a critical mass for investment and service delivery. The basin scale links upstream and downstream communities.

Conflicting objectives compromise long-term management

An objective gap occurs when conflicting objectives compromise long-term targets for urban water policy. This can happen because of diverging interests between water-related fields or political discontinuity (e.g. for short-term mayoral mandates; as an illustration, local mandates last three years in Mexico).

Water-related policy making involves a range of actors at the institutional, functional and hydrological scales, making the role of cities quite diverse within and across countries. Water conflicts between different municipalities in a given metropolitan area are increasingly common; governance is key to aligning interests. The cities surveyed signalled a lack of institutional incentives for co-operation (60%), contradiction between levels of governments (60%), competition between authorities (53%) and the interference of lobbies (50%) as major challenges to long-term urban water management (Figure 5.3).

Figure 5.3. **Perceived challenges to long-term urban water management**

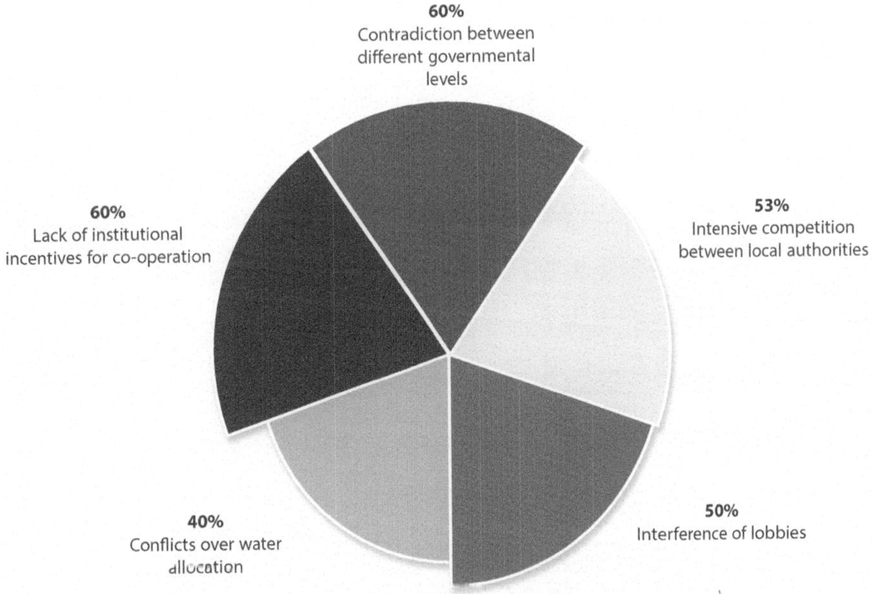

Note: results based on a sample of 30 respondents who indicated the options provided represent an obstacle ("major", "important" and "somewhat an obstacle").

Source: OECD (forthcoming b), *Water Governance in Cities*, OECD Publishing, Paris.

Institutional and policy silos

Water management is fragmented, and the responsible institutions at city, national or other levels tend to work in silos. As an example, several line ministries and other governmental agencies are involved in regulating water services, including environment ministries (for managing water pollution); health ministries (for setting and monitoring water quality standards) and economics and finance ministries (for tariff regulation) (OECD, forthcoming a). At the vertical level, water services are characterised by multi-level governance from supra- to sub-national levels. The European Union provides an illustration of supra-national regulatory powers in the water service sector, notably through the EU Water Directive. At sub-national level, municipalities are generally responsible for providing and managing water and wastewater service delivery.

Grey areas or duplications in the allocation of roles and responsibilities can contribute to poor water governance if co-ordination mechanisms are not in place to foster systemic approaches and trigger political commitment to co-operating at all levels. The cities surveyed reported the fragmentation of water-related tasks (50%), overlapping roles and responsibilities (47%), and the lack of co-ordinated legislation (40%) as prominent obstacles to policy coherence. The lack of co-ordinated legislation reflects situations (for example) where one piece of legislation that may have negative consequences on water quality can take precedence over other legal dispositions meant to prevent or remedy these potential negative consequences.

The cities reported several areas that directly influence water policy (Figure 5.4) and use a range of co-ordination mechanisms to ensure policy coherence (Box 5.1).

- Spatial planning (87%) influences the way water is managed within the city and can contribute to integrated water resource management and water security.
- Energy (83%) and water are closely connected: wastewater can generate energy, and energy is often required to tap alternative water sources, e.g. investment in desalination can enhance water security at the expense of energy security (as in Barcelona). Policy coherence can reduce cities' energy input and heat output while promoting efficient water uses.

Figure 5.4. **Policy areas influencing water governance in cities**

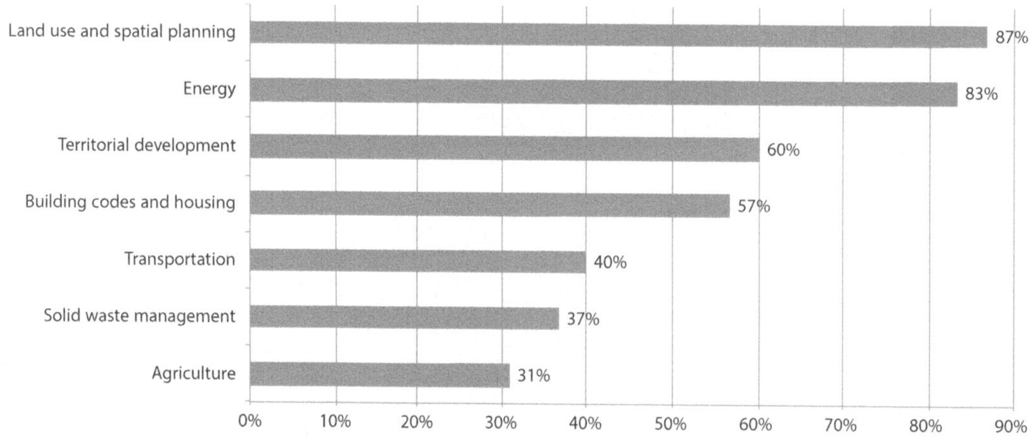

Note: results based on 30 respondents that indicated the influence from policy areas as "critical" and "important".

Source: OECD (forthcoming b), *Water Governance in Cities*, OECD Publishing, Paris.

- Building codes and housing (57%) aim increasingly to protect citizens from rising sea levels and extreme weather while ensuring increased water efficiency.
- Solid waste can impair water quality and quantity (37%) when discharged into water bodies.
- Agriculture raises urban water governance challenges (31%), especially in terms of controlling and avoiding the harmful consequences of fertiliser and pesticide use.

Box 5.1. **Co-ordinating water and related policies**

Water and spatial planning: **Cologne** co-ordinates water and spatial planning for new building areas to prevent flood damages due to heavy rainfalls; **Melbourne**'s Water Sensitive Urban Design integrates water cycle management in urban planning and design; **Mexico**'s *Desarrollos Urbanos Integrales Sustentables* integrates water and environmental aspects; **Glasgow**'s Metropolitan Strategic Drainage Partnership raises awareness of constraints to drainage infrastructures (and their resolution) and supports authorities in preparing new local and strategic planning policies. Different examples of urban-flood prevention through land use planning also exist: integrated strategic planning in **Paris** provides for better integration of stormwater management in urban decision-making processes; **New York** and **London** also promote strategic plans aiming to maximise the use of green infrastructure and other urban facilities to avoid urban floods; **Seoul** has attempted a number of initiatives, such as building 77 additional drainage facilities able to retain 554 054 square metres of water, but there is room to improve urban planning.

Water and energy: they are inextricably linked to each other. Energy is required to extract, treat and transport water, and it takes water to produce energy – especially for cooling steam-electric power plants. Municipalities spend between 25% and 60% of their budgets to supply energy for their water infrastructure. **Budapest** enforces legal requirements for co-ordination between the water supply and energy sectors; **Copenhagen** has completed several water-saving campaigns (e.g. "Max 100") to raise citizens' awareness of their daily water consumption.

Water and agriculture: the use of fertilisers and pesticides can have harmful consequences on water. Local management initiatives and collective arrangements, including financial transfers for tackling the issue of water competition, are in place. Decentralised urban-rural arrangements aim to reduce water pollution from agriculture, while agreements with farmers endeavour to reduce flood risk upstream.

Water and regional development: relevant policies for water management go beyond the water sector and may concern other policies (e.g. land use and transport planning). On the one hand, water is essentially a local issue (it is pumped, treated, distributed and used locally). On the other hand, water has implications beyond the local scale, as it drives or hinders regional economic development, competitiveness and assets within the national framework. Given the importance of local actors and territorial specificities in the water sector, policy makers should find ways to maintain coherence while preserving diversity.

Water and environment: published in 2002, **London**'s Biodiversity Strategy recognises the importance of the River Thames and other waterways for biodiversity and promotes the restoration of degraded tributary rivers. **New York**'s Department of Environmental Protection is leading the Green Infrastructure Program to promote, among others, the natural movement of water and manage stormwater run-off from streets, sidewalks, parking lots and rooftops.

Financial bottlenecks to effective governance

A funding gap occurs when unstable or insufficient revenues undermine effective implementation of water responsibilities at sub-national level, hinder cross-sectoral policies and investments for water infrastructure (OECD, 2011). The funding gap has different origins; some of which are already discussed in Chapter 2. The cities surveyed reported difficulties in raising tariffs for water services (67%); affordability constraints requiring tariff adjustments (67%); and weak prioritisation of investments (43%) as major obstacles to their capacity to manage water effectively (Figure 5.5). Additionally, 37 % of the cities surveyed lack multi-annual strategic plans and budgets, while 37% do not have financial guarantees to borrow money. In some cases (23% of respondents), a "golden rule" on public finance restricts cities' borrowing capacity; 1 in 5 cities (20%) also reported difficulties in mobilising the private sector.

Figure 5.5. **Perceived challenges to the financial sustainability of urban water management**

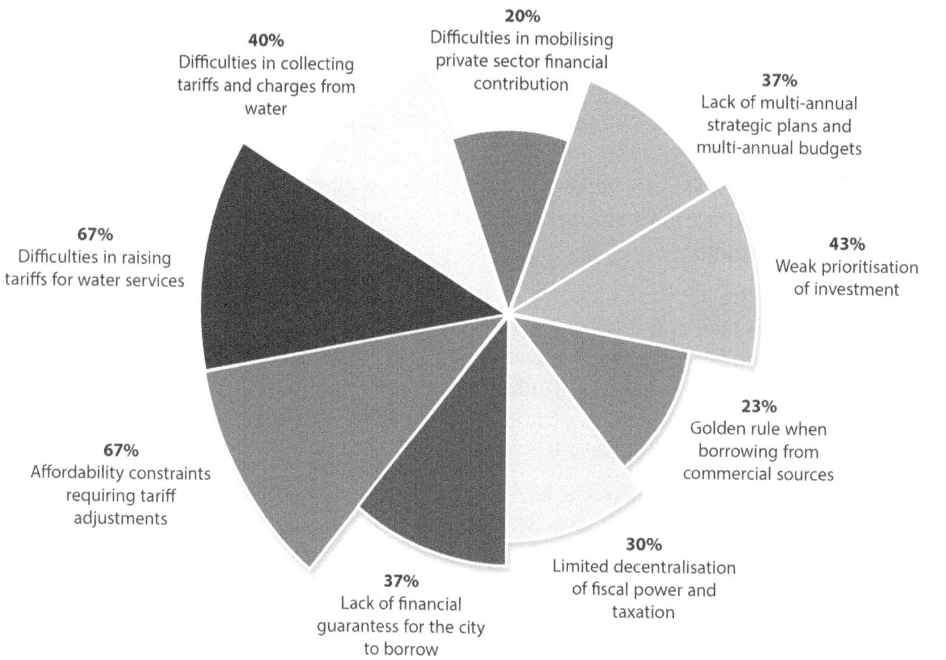

Note: results based on a sample of 30 respondents who indicated the options provided represent an obstacle ("major", "important" and "somewhat an obstacle").

Source: OECD (forthcoming b), *Water Governance in Cities*, OECD Publishing, Paris.

Almost all cities surveyed have established retail water tariffs or user charges. Most cities use additional economic instruments – most commonly fines, water pollution charges and levies – for water management (Figure 5.6). In addition, 70% of cities surveyed play a role in tariff regulation. For example, Stockholm's governing board sets the principle for defining tariffs: they have to be affordable, sustainable for financing water management and encourage local stormwater treatment. The tariffs are used to produce drinking water, maintain the water and sewage network, and treat the wastewater from buildings and stormwater on the roads. City councils in Calgary and Montreal and French cities also set tariff rates, and Milan's Autorità Territoriale Ottimale (ATO) sets the tariff to be approved by the regulator, the National Electricity and Gas Authority. In metropolitan Copenhagen – which numbers eight municipalities – the multi-supply company HOFOR is responsible for setting the tariff, subject to a nationally set price cap.

Figure 5.6. **Use of economic instruments for urban water management in OECD countries**

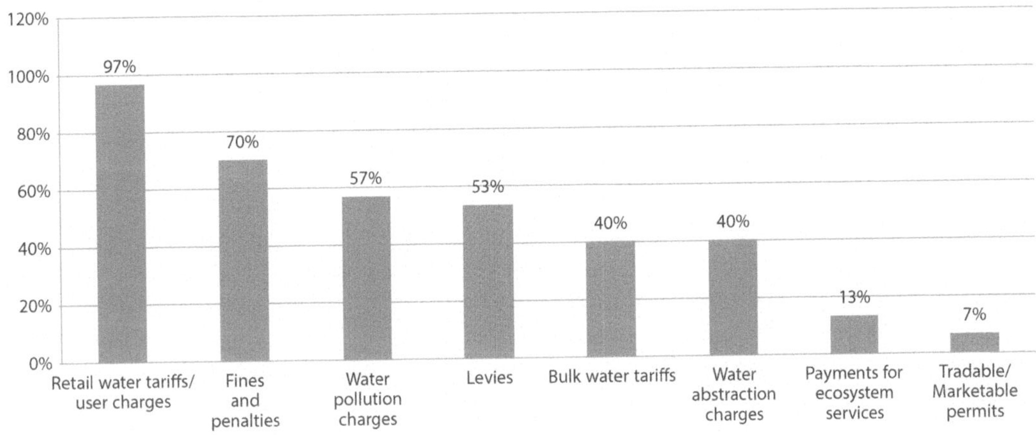

Note: Results based on a sample of 14 respondents who responded "yes" to the options provided.

Source: OECD (forthcoming b), *Water Governance in Cities*, OECD Publishing, Paris.

The role of central governments in tariff-setting reflects the degree of decentralisation in the country. In Mexico, each municipality sets its tariff structures according to the laws of each federal state; the federal level has limited powers to influence the process – it can only do so by establishing a voluntary norm and providing financial incentives as part of federal programmes (OECD, 2013). In the Czech Republic, individual utilities set the tariffs, subject however to price controls by the finance ministry. In some countries (e.g. Portugal – see the section on dedicated regulatory bodies), the national level defines the basic principles for tariff structures and levels – including the number of blocks – while leaving some degree of flexibility to factor in local, technical and social conditions.

Many cities in OECD countries have introduced measures to make water more affordable to the population at large and to selected groups. These measures usually target poor populations (70% of cities surveyed), populations living in slums (37%), ethnic minorities (30%) and disabled persons (43%). Such measures include reducing the value added tax or wastewater tax, using progressive social tariffs, avoiding water disconnection, setting up a national solidarity fund and abolishing annual fixed fees.

Capacity: The Achilles' heel of sub-national governments

Local actors' lack of scientific, technical and infrastructural capacity hinders the design and implementation of water policies or strategies within cities and beyond. The capacity gap has spillover effects: it often triggers an information gap (in terms of quantity, quality, type), which in turn can generate an accountability gap. Many cities in OECD countries face serious capacity challenges to manage water in the face of future challenges, particularly when utilities are operating in the red; when increasingly stringent environmental regulation cannot be enforced at the lower levels; when access to technological innovations is too costly; when the water sector does not attract sufficient professionals (e.g. the Netherlands foresees a shortage of 20 000 water professionals in the next 2 decades); or when systems are not in place to produce, use and share policy-relevant data for decision-making and transparency.

The cities surveyed reported the lack of staff and managerial competences (63%) as the main source of capacity gap, followed by the difficulty of ex post monitoring and evaluation (40%) and poor planning often not articulated with national legislation (37% of cases) (Figure 5.7). In Mexico, funds allocated by federal programmes to sub-national governments cannot always be disbursed and sometimes have to be returned (with penalties) because of limited local capacity to develop and implement good projects. The professionalisation of water staff is also a key challenge for capacity-building at the sub-national level.

The capacity to produce, collect and share quality data varies from one city to another. Depending on the purpose, data might be collected by the local authority, the service provider, statistical offices or environmental agencies. Key issues concerning data include the *what* (available information is too technical – 53% of respondents), the *when* (data collection is incomplete and irregular – 47% of respondents) and the *how* (data is dispersed across agencies, making it difficult to track and compare – 47% of respondents) (Figure 5.8). Overly technical data hinders participation by lay people. For example, the complexity of London's water planning system, including the highly technical modelling that feeds the figures in the plans, makes it very difficult for non-technical stakeholders to contribute and hinders comparisons over time (Greater London Authority, 2011).

Figure 5.7. **Capacity challenges**

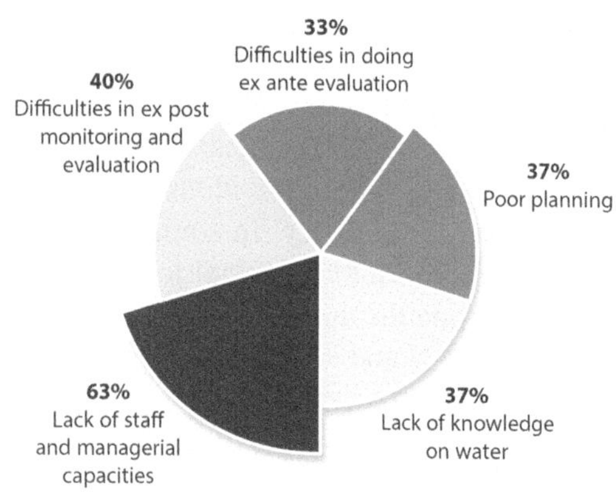

Figure 5.8. **Information challenges**

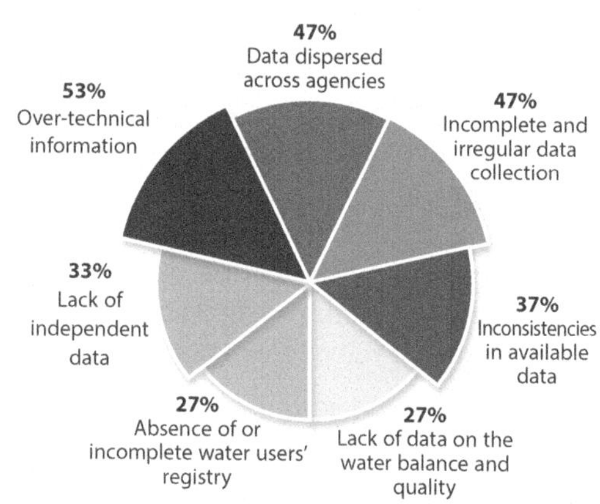

Note: results based on a sample of 30 respondents who indicated the options provided represent an obstacle ("major", "important" and "somewhat an obstacle").

Source: OECD (forthcoming b), *Water Governance in Cities*, OECD Publishing, Paris.

Furthermore, the water sector displays important information asymmetries – which, when applied to the status of water supply and sanitation (WSS) assets, may lead to sub-optimal contractual arrangements between the authority and the service provider, market abuse and consumer mistrust. Information asymmetries were reported as the most important obstacle (53%) to the accountability of water decision-makers (Figure 5.9). Easy and transparent access to data could address this type of challenge – which, in addition to monopolistic behaviours, justifies resorting to regulatory instruments to protect the public interest in tariff-setting and investment decisions (OECD, forthcoming a).

Accountability

The cities surveyed reported a range of concerns that hinder the transparency and participatory nature or urban water management (Figure 5.9), including: limited information-sharing across local authorities (53%); the lack of benchmarking to evaluate water quality and quantity, as well as service providers' performance (47%); and limited monitoring to guide decision-making (40%). These gaps may also result from a lack of human and financial resources and expertise.

Figure 5.9. **Perceived challenges to transparency and accountability for urban water management**

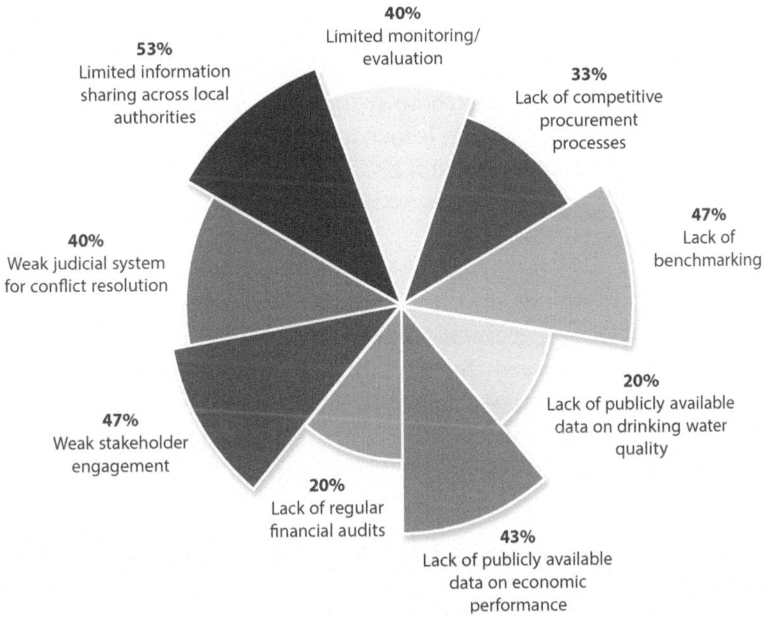

Note: results based on a sample of 30 respondents who indicated the options provided represent an obstacle ("major", "important" and "somewhat an obstacle").

Source: OECD (forthcoming b), *Water Governance in Cities*, OECD Publishing, Paris.

Figure 5.10. **Established mechanisms for assessing the performance of urban water management**

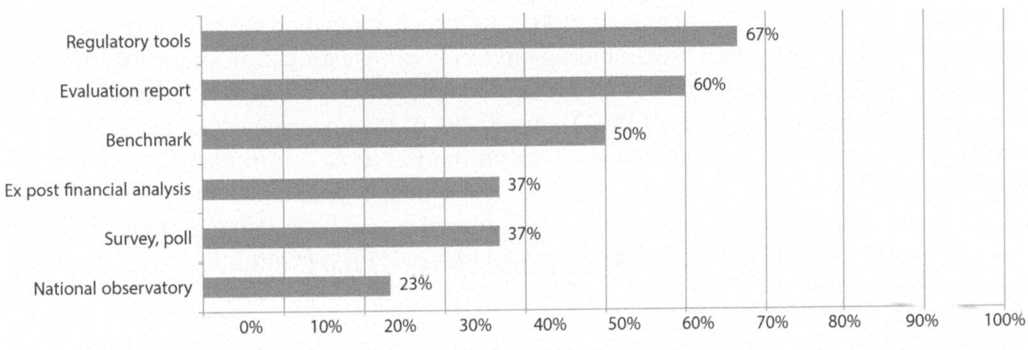

Note: results based on a sample of 30 respondents who indicated the mechanisms were used "very often" and "often".

Source: OECD (forthcoming b), *Water Governance in Cities*, OECD Publishing, Paris.

A range of mechanisms can be used to assess the performance of water management in cities with a view to increasing transparency and accountability (Figure 5.10). Benchmarks and evaluation reports measure performance against indicators. Benchmarking allows competition-by-comparison to achieve transparency, while evaluation provides citizens with useful information to guarantee accountability. Financial analyses help draw a clear picture of local government's financial needs. National observatories can monitor service delivery performance and improve information transparency and the accountability of the water service sector. The section below covers the role of regulation.

Governance arrangements for urban water management

Coping with multi-level governance gaps for urban water management requires an institutional setting and tools that allow strengthening co-ordination across people, places and policies, and addressing the market failures identified above. This section explores some of the governance arrangements (both institutions and tools) that help address the seven governance gaps mentioned above. It focuses on metropolitan governance, dedicated regulatory bodies for water utilities and stakeholder engagement.

Metropolitan governance

Rapid socio-economic trends are pushing towards greater co-ordination across municipalities to jointly face investment choices and policy decisions. A recent OECD study (Kim, Schumann and Ahrend, forthcoming), pointed out that the pressing need to build more effective metropolitan governance for stronger, more inclusive and sustainable growth is all the more salient in a context of recent crises and long-term pressure on public finances. The study categorises governance arrangements in four groups (Box 5.2 provides an illustration in the water sector):

- **informal/soft co-ordination**, lightly institutionalised platforms for information-sharing and consultation
- **inter-municipal authorities**, sharing costs and responsibilities across member municipalities
- **supra-municipal authorities**, as an additional layer above municipalities, introduced either by creating a directly-elected metropolitan government, or by upper governments setting down a non-elected metropolitan structure
- **"metropolitan cities"**, a special status for large cities that puts them on the same footing as the next upper level of government and gives them broad competences.

These governance arrangements are likely to affect water management arrangements between the urban core and its surroundings in terms of investment, information-sharing, monitoring, stakeholder engagement and policy complementarities across different sectors (see Box 5.2). According to the OECD survey on water governance in cities (OECD, forthcoming b), a wide range of informal co-ordination tools are in place beyond inter-municipal co-ordination (listed by 40% of respondents), including shared databases/ information systems (53%); performance indicators (47%); joint financing projects (47%); and platforms for dialogue between sub-national players (40%) (Figure 5.11).

Box 5.2. **Metropolitan arrangements in the water sector**

Informal/soft co-ordination consists of lightly institutionalised platforms for water-related information-sharing and projects:

- The Metropolitan Glasgow Strategic Drainage Partnership (MGSDP) is a collaborative venture between local authorities (led by the Glasgow City Council), the Scottish Environment Protection Agency, Scottish Water, Scottish Enterprise, Clyde Gateway and Scottish Canals. It was established to upgrade the Glasgow area's drainage and sewerage network, reduce flooding and support urban development requirements while improving water quality and the environment.

Supra-municipal co-operation: an additional layer above municipalities, introduced either by creating a directly-elected metropolitan government, or by upper governments setting down a non-elected metropolitan structure. The co-operation across municipalities for water is based on established joint organisations (e.g. companies or authorities) with co-ordination roles over water policy, planning and service provision:

- *Conseil communautaire* (France): this elected body acts on behalf of the municipalities on specific water issues (i.e. water allocation; drinking water provision; research; operation/maintenance of infrastructures).
- Barcelona Metropolitan Area (BMA, Spain): the metropolitan authority has responsibilities throughout the water cycle, from the drinking water supply to wastewater treatment and reclaimed water production. It encompasses 9 utilities across the metropolitan area's 36 municipalities, 7 wastewater treatment plants and 3 reclaimed water plants. Managing urban waters at the metropolitan level has fostered a wider perspective at the water cycle level, as well as shared infrastructure and expenses. The BMA encourages customer involvement to learn about different territorial needs and expectations. For the future, Barcelona is looking at alternative water resources and strengthened water cycle management. Aguas de Barcelona was created in 2013, jointly with a large metropolitan utility, to manage drinking, reclaimed and wastewater for all the metropolitan territory.
- Metro Vancouver: the regional government, operating under the name "Metro Vancouver", provides services through four corporate entities, one of which is the Greater Vancouver Water District. It has a role in watershed management, water treatment, water transmission, wholesale distribution to municipalities, monitoring and reporting on Metro Vancouver water quality, and planning for Metro Vancouver water system's sustainability.
- Authorities for Optimal Territorial Districts (Italy): these autonomous entities made up of municipalities are responsible for water planning, water resource management and identifying service providers. They cover the area of the province or river basin.

Inter-municipal single-purpose co-operation on water: contracts or ad hoc agency for providing water service across municipalities aim to share costs and responsibilities (sometimes with the participation of other levels of government and sectoral organisations):

- Horizontal co-operation:
  - Association: the Water Management Association of the West Bohemia Region in the Czech Republic, a voluntary union of 91 municipalities and 2 associations of municipalities from 5 districts (Karlovy Vary, Tachov, Sokolov, Chomutov and Rakovník), was established in 1993 to manage, operate and develop water supply and wastewater treatment systems – which is a legal obligation of municipalities in the country. The number of association members has more than doubled from its initial 40 municipalities in the Karlovy Vary region and now covers a population of about 186 000 inhabitants. Member municipalities transfer their water assets to

Box 5.2. **Metropolitan arrangements in the water sector** *(continued)*

this voluntary association upon joining it and take them back when leaving. The governing body of the association is the general meeting, where each municipality has one vote.

- *Syndicats intercommunaux* (France): these intercommunal syndicates are run by joint committees, representing members of each local council and levying a compulsory contribution for water supply.
- *Mancomunidades* (Spain): this administrative form is meant for purely inter-municipal co-operation, in which municipalities appoint local politicians to the governing body of the *mancomunidades*. The number of appointees is proportional to the size of the respective member municipalities' population. *Mancomunidades* help to reach scale economies for water services.

• Vertical co-operation:

- Consortia (Italy and Spain): these standing organisations have a board and staff to manage the drinking water supply cycle, from production to distribution (e.g. the Greater Bilbao Water Partnership, a consortium of 43 municipalities, the provincial government of Biscay, the Autonomous Basque Community and central government).

Metropolitan cities: cities that exceed a legally defined population threshold can be upgraded to a special status as "metropolitan cities", putting them on the same footing as the next upper level of government. They have broader competences in the water sector, including planning, policy making, strategy-setting and service provision.

• The Seoul Metropolitan government sets water management policies and takes actions to improve water quality through an online monitoring system for water quality.

• The Greater London Authority of the Mayor of London and the 25-member London Assembly and is also responsible for strategic local government in the water sector.

Source: OECD (2015), *Governing the City*, OECD Publishing, Paris, http://dx.doi.org/10.1787/9789264226500-en; Kim, S.-J., Schumann A. and Ahrend R. (forthcoming), *What Governance for Metropolitan Areas?*, OECD Regional Development Working Papers, OECD Publishing, Paris; Kurki, V.O., T.S. Katko and E.P. Pietilä, "Bilateral Collaboration in Municipal Water and Wastewater Services in Finland", *Water 2010*, Vol. 2, pp. 815-825; United Nations Development Programme, *Inter-municipal co-operation*, www.municipal-cooperation.org/index.php?title=Czech_Republic; www.municipalcooperation.org/images/5/5e/Paper_Institutional_Shifts_in_IMC_Service_Delivery_Spain_2007.pdf; www.services.eaufrance.fr/docs/synthese/rapports/spea2009_201202_EN.pdf.

Metropolitan authorities have gained increasing competences in water management in the last decade. When they exist, they mainly operate as policy facilitators, favouring information exchange across municipalities in the metropolitan area (Figure 5.12). Most also have competences in strategic service provision management (e.g. setting performance targets, hiring senior managers, organising calls for tenders and supervising sub-contractors), as well as legislative, financial or technical competences. Less than a third provide operational management for the drinking water supply and sanitation services; in the OECD survey, Veracruz, Acapulco, Barcelona and Nantes are exceptions. The case studies of Auckland, Tokyo Metropolitan Government and Korean cities featured in this report provide detailed illustrations of the organisation and operation of supra-municipal water management bodies.

Figure 5.11. **Mechanisms for co-ordinating water policy across government levels**

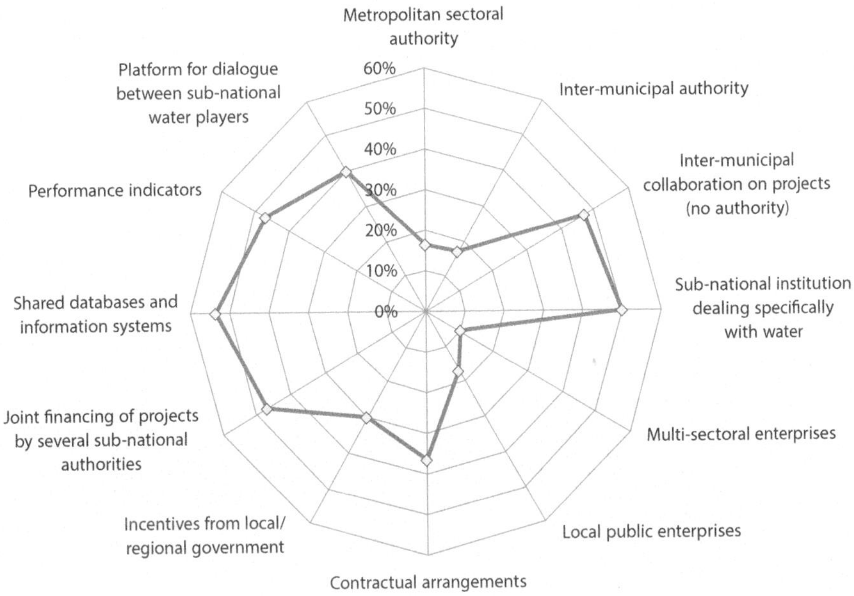

Note: Results based on a sample of 14 respondents who responded "yes" to the options provided.

Source: OECD (forthcoming b), *Water Governance in Cities*, OECD Publishing, Paris.

Figure 5.12. **Water competencies of metropolitan bodies**

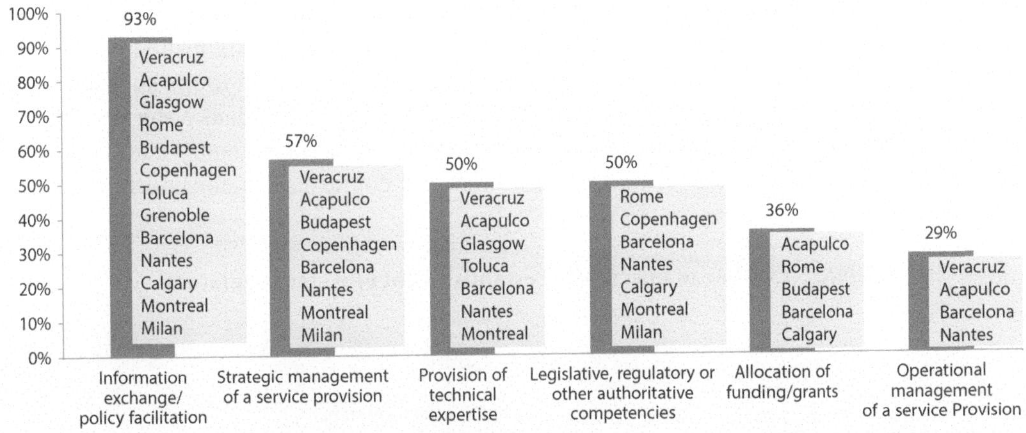

Note: Results based on a sample of 14 respondents who responded "yes" to the options provided.

Source: OECD (forthcoming b), *Water Governance in Cities*, OECD Publishing, Paris.

Strategies for metropolitan-level water governance may offer interesting models for application to the sector, e.g. increasing co-ordination across functional and hydrological logics for integrated urban water management; pooling resources for water resources management and service delivery; and increasing policy coherence and creating synergies across sectors.

Establishing dedicated regulatory bodies

In response to the governance gaps identified above, a large spectrum of regulatory functions need to be performed in relation to urban/networked water services. These regulatory functions can be economic or environmental, or embrace social issues such as equity, affordability and universal coverage. They are summarised in a typology developed by the OECD (see Annex 5.A3).

Regulatory functions do not necessarily have to be in the hands of a single responsible institution. However, recent evidence across a number of OECD countries points to the consistent establishment of dedicated regulatory bodies in charge of overseeing networked water services (i.e. overseeing urban water services and centralised systems over a certain size).

This specific institutional mechanism is seen as responding to the governance gaps identified earlier in the chapter, particularly the need to protect the public interest in a sector that is both complex and prone to market failures. This section analyses recent trends in this important area that contribute to shaping urban water management, building on the OECD survey on the Governance of Water Regulators (OECD, forthcoming a).

The emergence of dedicated water regulators

Countries have adopted different institutional organisations to ensure that the various regulatory functions related to water services are performed. A large literature reviews the different regulatory models for services. Among the various institutional organisations, the development of dedicated regulatory bodies stands out across countries as a growing response to some of the pitfalls of regulatory frameworks for water services.

The OECD survey on the Governance of Water Regulators (OECD, forthcoming a) documents the consistent emergence of dedicated regulatory bodies for water services over the past two decades. While North American water regulators were established before the Second World War, most of the other water regulators who responded to the survey came into force between 1990 and 2009. Over the past four years, six additional water regulators became operational, mainly in Europe (Belgium/Flanders, Estonia, Hungary and Italy), but also in Victoria, Australia (Figure 5.13). Today, several countries are considering setting up a water regulator.

Figure 5.13. **Year of operational establishment of water regulators**

Number of regulators/34

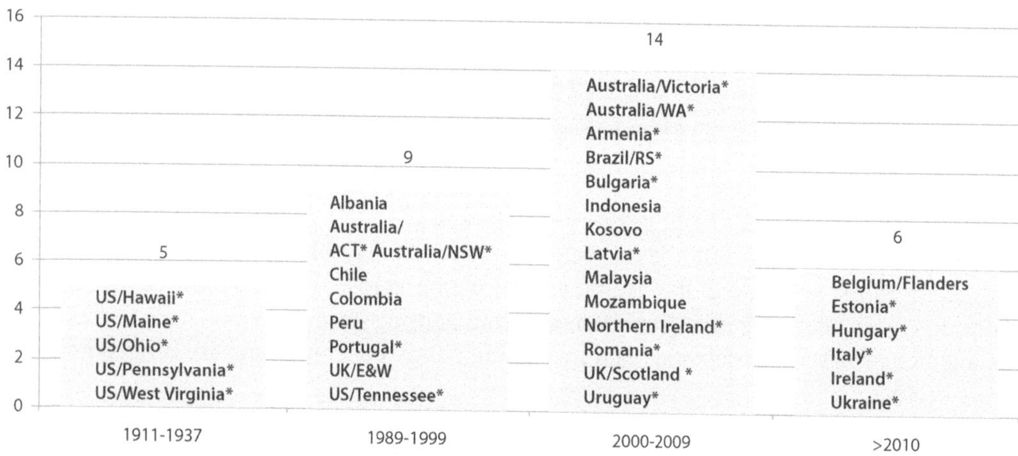

Note: *multi-sector regulators.

Source: OECD (forthcoming a), *The Governance of Water Regulators*, OECD Publishing, Paris.

Why are they established?

The vast majority of regulators who answered the OECD survey justify their establishment on the basis of protecting the public interest (Figure 5.14); other prominent motivations are making service providers more accountable and being part of a broader process of regulatory reform. In a number of cases, establishing a regulator is also justified by the need to make price-setting an independent process (i.e. at arm's length from government and protected from capture by private interests) or catalyse regulatory expertise in the public sector.

Figure 5.14. **Main justifications for establishing a water regulator**

Number of regulators/33

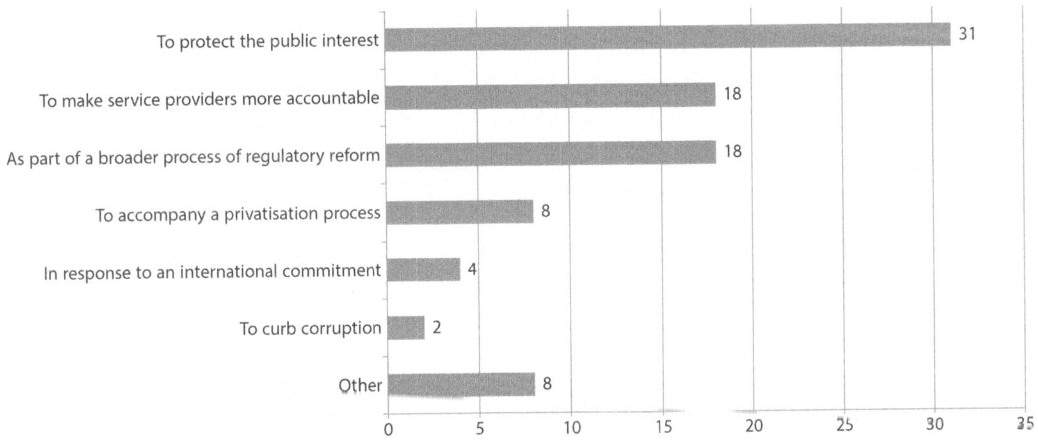

Source: OECD (forthcoming a), *The Governance of Water Regulators*, OECD Publishing, Paris.

Regulators responsible for water services do not work in isolation. They are part of a broader regulatory framework at the national and sub-national levels that typically involves line ministries (environment or natural resources) and various public agencies. Agencies in charge of environmental protection co-ordinate with water regulators to monitor the environmental sustainability of WSS services. Local governments can be involved in WSS regulation: in Portugal, municipalities determine retail tariffs in collaboration with the water regulator as a shared competence.

What do water regulators do?

Regulators mainly oversee urban activities and centralised systems above a certain size. In several cases – e.g. the Australian Capital Territory and Chile – the regulator does not have the prerogative for rural water at all. The establishment of water regulators has usually gone hand in hand with deep reforms of the water industry – particularly the corporatisation of water operators (in the Australian Capital Territory) and the consolidation of water service provision around fewer but bigger providers (in Ireland and Portugal).

Based on the survey, water regulators play a critical role in economic regulation, performance monitoring, regulatory enforcement and customer engagement. This positions them as critical actors in reforms aiming to establish more transparency and a user-centric approach. Regulators also constitute a vital link in the regulatory-governance cycle to support concrete implementation of government policies by ensuring the compliance and credibility of the regulatory framework.

- **Economic regulation (tariff-setting and review of utilities' investment plans)**: all the regulators surveyed (except one) have responsibility for tariff regulation; in a few cases, they have a mainly advisory role (e.g. in Belgium/Flanders and Hungary).
- **Data collection and performance monitoring related to water services**: all but two of the regulators surveyed have responsibility for monitoring service delivery performance, often in combination with information and data gathering; most regulators have the power to collect information from regulated entities through compulsory processes.
- **Enforcement of regulations and standards**: regulators consistently have strong enforcement roles and powers. Most regulators can enforce compliance with regulation, investigate breaches and issue codes of conduct and guidelines, either independently or together with other bodies; most also have the power to impose fines and financial sanctions against regulatory infringements.
- **Customer engagement and protection**: more than two-thirds of regulators play a role in customer engagement and protection.

Based on the results of the OECD survey on the Governance of Water Regulators (OECD, forthcoming a), water regulators display a strong culture of consultation. Two-thirds have a legislative requirement to consult with regulated entities before making a regulatory determination and to conduct public consultation in advance of making a regulatory decision. When not required by law, they routinely carry out consultation on a voluntary basis.

Regulators are subject to strong legislative reporting requirements. Most are required to report to the legislature on their performance (usually on an annual basis) and to publish information on the various dimensions of their activity, particularly their operating costs, decisions, resolutions, agreements and governance structure. All but one must produce an annual report detailing their activities.

Stakeholder engagement

Stakeholder engagement is herein defined as the process through which individuals, groups and organisations have the opportunity to participate in decision-making that will affect them, or in which they have an interest. Stakeholder engagement is key to co-ordinating various urban actors and interests. As a governance instrument, it can help build trust and ownership, secure willingness to pay for water services, raise awareness on current and future water challenges, ensure the accountability of city managers and service providers to end users and citizens, manage conflicts on water allocation, ensure the political acceptability of different ownership models and set convergent objectives across policy areas. Stakeholder engagement mechanisms can bring together urban planners, water service providers, regulators, advisers and civil society to develop dynamic integrated approaches.

City departments interact with various authorities to manage water (Figure 5.15). Their main counterparts are service providers (listed by 46% of respondents to the OECD survey on water governance in cities [OECD, forthcoming b]), followed by regional governments (27%), local governments (23%) and customers and their associations (23%). The interaction with central governments is less frequent (it takes place "sometimes" for 33% of respondents and "never or rarely" for 17% of respondents). They also have rather low interaction with irrigators and their associations (50% of cities surveyed point out that they *never* interact with this type of stakeholder), civil society (40% "sometimes" interact with this stakeholder) and business/industry (40% "sometimes" interact with this stakeholder).

Figure 5.15. **Frequent interactions between cities and stakeholders**

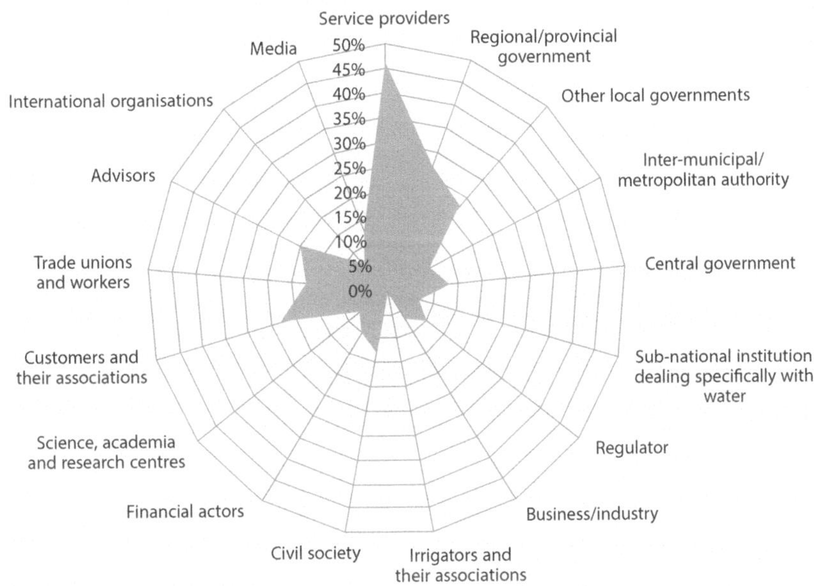

Note: results based on a sample of 30 respondents who indicated the interactions to occur "always, very frequently".

Source: OECD (forthcoming b), *Water Governance in Cities*, OECD Publishing, Paris.

Technology plays an important role in engaging stakeholders: 77% of respondents use web-based communication technologies (online platforms, email, social media, website and apps) on a regular basis to engage with stakeholders; 67% use traditional media (newspapers, newsletters, TV, radio, etc.); and more than 60 % hold regular meetings to consult and engage in water-related decision-making.

Stakeholder engagement in water governance has been largely incentivised in the broader context of a bottom-up call for open government. However, several practical obstacles are noteworthy (Figure 5.16). A full 50% of surveyed cities state that the major obstacle is the complexity of the issues at hand and resistance to change. Other challenges are the lack of clarity and feedback on the expected use of stakeholder inputs (leading to consultation fatigue); consultation capture from over-represented categories; the absence of political will and leadership; lack of time, staff and funding; weak supportive legal frameworks; weak capacity; lack of citizen concern and awareness; information asymmetry; and fragmented settings.

The principles deriving from OECD work on stakeholder engagement for inclusive water governance (Box 5.3) provide some guidance that cities may wish to consider to overcome obstacles to stakeholder engagement in water governance. Central governments have a role to play in promoting and facilitating the effective implementation of these principles.

Figure 5.16. **Obstacles to effective stakeholder engagement in urban water management**

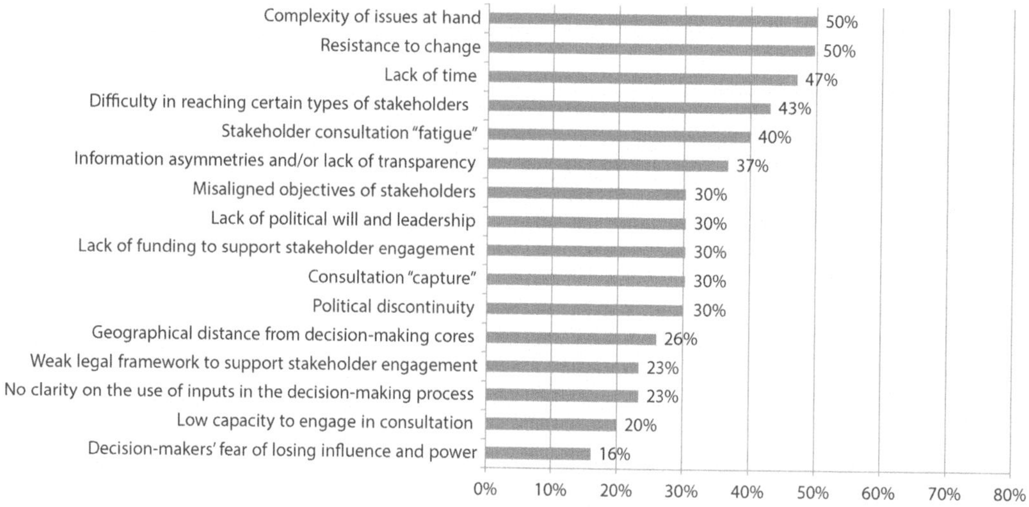

Note: results based on a sample of 30 respondents who indicated the obstacles as being "critical" and "important".

Source: OECD (forthcoming b), *Water Governance in Cities*, OECD Publishing, Paris.

Box 5.3. **Principles for effective stakeholder engagement in water-related decision-making**

- Inclusiveness and equity: map all those who have a stake in, or are likely to be affected by, the outcome. List their responsibilities, core motivations and interactions. A stakeholder analysis would help identify stakeholders, their interests and their potential role in supporting the decision-making process in cities. Stakeholder analysis can help understand power relations. Stakeholder engagement increases the effectiveness and acceptability of policies, demonstrating the benefit of shifting away from top-down management processes run exclusively by professionals and governmental authorities.
- Clarity, transparency and accountability: define the ultimate decision-making line, stakeholder engagement objectives and expected input use. Stakeholders need to be engaged in different phases of the decision-making process and provided with the information needed for successful interaction. Their involvement helps make the process more transparent, as well as overcome politicians' short-termism and election interests.
- Capacity and information: allocate proper financial and human resources, and share needed information for result-oriented stakeholder engagement. Urban water management involves a plethora of complex issues. Dealing with them might be difficult and generate "stakeholder fatigue". It is important to keep the motivation high, share visions and communicate in an understandable language. Involving unheard voices in the process helps overcome the technical approach of experts in the field, and implies investing in education and training to building awareness and promote interaction.
- Efficiency and effectiveness: regularly assess the process and outcomes of stakeholder engagement, and learn, adjust and improve accordingly. The city needs to ensure that the inputs from stakeholder engagement are taken into account and that the outcomes are regularly assessed with a view to improving the decision-making process.

Box 5.3. **Principles for effective stakeholder engagement in water-related decision-making** *(continued)*

- Institutionalisation, structuring and integration: embed engagement processes in clear legal and policy frameworks, organisational structures/principles and responsible authorities. The water issues at stake have consequences not only for the city itself, but for the surrounding area as well. Engaging stakeholders at the right scale is vital to scaling up from the local level to the city level and then the metropolitan level when required. Institutionalising the process is crucial to strengthening formal participation.
- Adaptation: customise the type and level of engagement to the needs and keep the process flexible and adaptable to changing circumstances. Stakeholders and authorities in charge of urban water need to be flexible and adapt to the context. This implies overcoming resistance to change and adopting a long-term vision allowing for shifting perspectives on what is relevant not only for present generations, but also for future generations.

Source: adapted from OECD (forthcoming c), *Stakeholder Engagement for Inclusive Water Governance*, OECD Publishing, Paris.

Note

1. http://stats.oecd.org/glossary/detail.asp?ID=4497.

Annex 5.A1

Characterisation of respondents to the OECD Survey on urban water governance

Cities	Size[a]	Speed of urbanisation	Spatial patterns	Governance arrangements[b]
Hong Kong Singapore			N/A	N/A
Mexico City			N/A	
New York				N/A
Barcelona				
Cologne Milan Rome Stockholm				N/A
Montreal			N/A	
Athens Budapest Copenhagen				N/A
Phoenix				
Amsterdam				
Toluca			N/A	
Calgary			N/A	

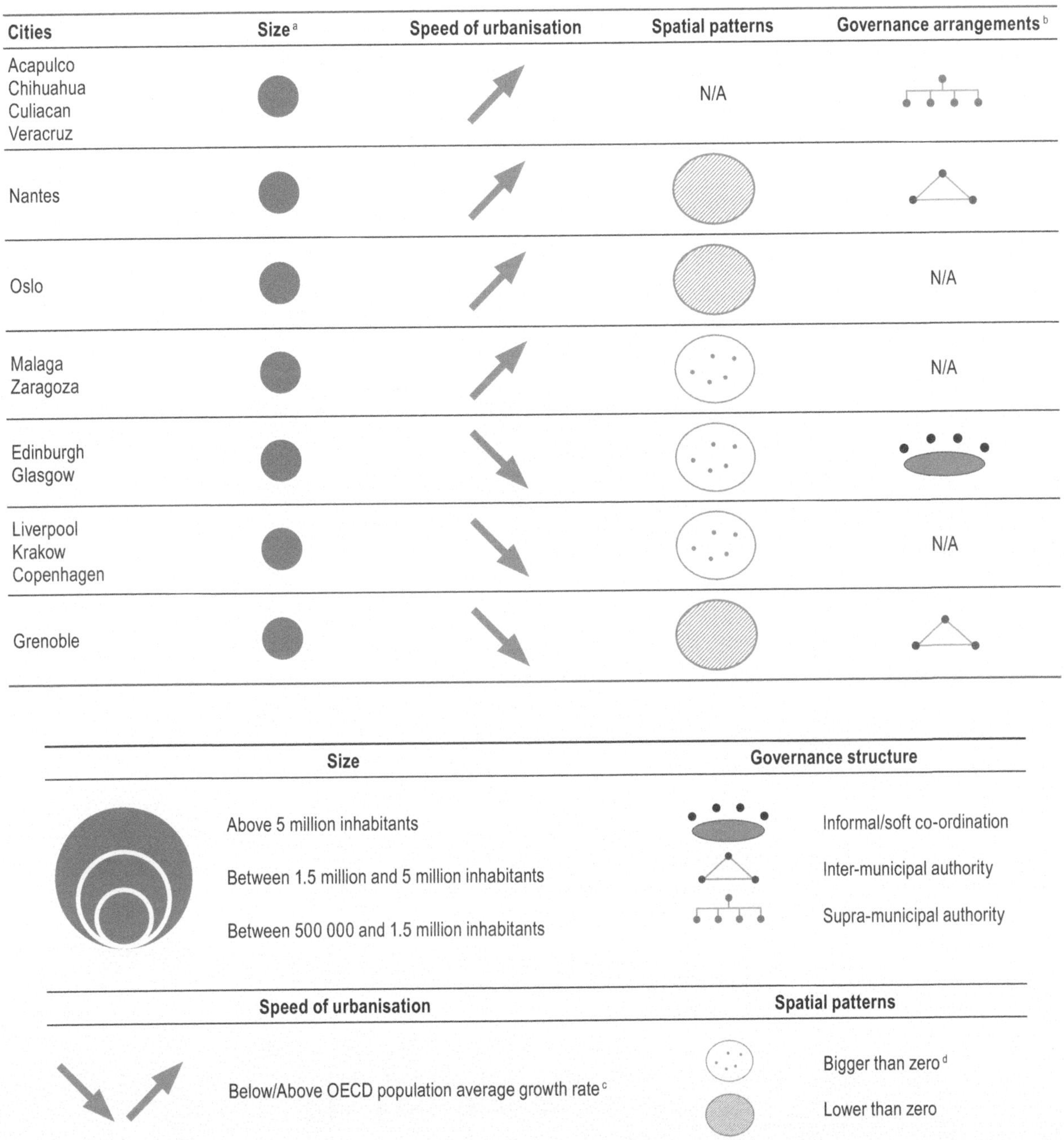

a. Total population metro area, 2012.

b. These governance arrangements are not specific to the water sector, but reflect the approach to multi-level co-ordination of the city and its surroundings.

c. OECD population average growth rate was 0.87 over 2000-12.

d. Looking at land use, cities can be distinguished in terms of compactness and sprawling. The sprawl index (SI) allows classifying cities according to the measure of the growth in built-up area, adjusted for the growth in city population. When the city population changes, the index measures the increase in the built-up area relative to a benchmark where the built-up area would have increased in line with population growth. The SI is equal to zero when both population and built-up area are stable over time. It is bigger (lower) than zero when the growth of built-up area is greater (smaller) than the growth of population, i.e. the city density has decreased (increased) (OECD Metropolitan Database, http://stats.oecd.org/Index.aspx?Datasetcode=CITIES). Data refer to 2006.

Annex 5.A2

Regulators who responded to the OECD survey on the Governance of Water Regulators

	Country/territory	Name of the regulator
1	Albania	Water Regulatory Authority of Albania
2	Armenia	Public Services Regulatory Commission of the Republic of Armenia (PSRC)
3	Australia/Capital Territory (ACT)	Independent Competition and Regulatory Commission (ICRC)
4	Australia/New South Wales (NSW)	Independent Pricing and Regulatory Tribunal (IPART)
5	Australia/Victoria	Essential Services Commission
6	Australia/Western Australia (WA)	Economic Regulation Authority (ERAWA)
7	Belgium/Flanders	Water Regulator – Flemish Environment Agency
8	Brazil/Rio Grande do Sul (RGS)	Agência Estadual de Regulação dos Serviços Públicos, Delegados do Rio Grande do Sul (AGERGS)
9	Bulgaria	State Energy and Water Regulatory Commission (SEWRC)
10	Chile	Superintendencia de Servicio Sanitarios (SISS)
11	Colombia	Regulatory Commission for Water and Sanitation (CRA)
12	Estonia	Competition Authority
13	Hungary	Hungarian Energy and Public Utility Regulatory Authority
14	Indonesia	Jakarta Water Supply Regulatory Body
15	Ireland	Commission for Energy Regulation (CER)
16	Italy	Regulatory Authority for Electricity, Gas and Water (AEEGSI)
17	Kosovo	Water and Wastewater Regulatory Office of Kosovo (WWRO)
18	Latvia	Public Utilities Commission of Latvia (PUC)
19	Malaysia	National Water Services Commission
20	Mozambique	Water and Sanitation Regulatory Council (CRA)
21	Peru	Superintendencia Nacional de Servicios de Saneamiento (SUNASS)
22	Portugal	The Water and Waste Services Regulation Authority (ERSAR)
23	Romania	National Regulatory Authority for Municipal Services (ANRSC)
24	United Kingdom/England & Wales	Water Services Regulation Authority (Ofwat)
25	United Kingdom/Northern Ireland	Northern Ireland Authority for Utility Regulator (NIAUR)
26	United Kingdom/ Scotland	Water Industry Commission for Scotland (WISC)
27	Ukraine	National Commission of the State Public Utilities Regulation of Ukraine (SCWRM)
28	United States/Hawaii	Hawaii Public Utilities Commission
29	United States/Maine	Maine Public Utilities Commission
30	United States/Ohio	Public Utilities Commission of Ohio
31	United States/Pennsylvania	Pennsylvanian Public Utility Commission
32	United States/Tennessee	Tennessee Regulatory Authority
33	United States/West Virginia	Public Service Commission of West Virginia
34	Uruguay	Unidad Reguladora Servicios Energia y Agua (URSEA)

Source: OECD (forthcoming a), *The Governance of Water Regulators*, OECD Publishing, Paris.

Annex 5.A3

Typology of regulatory functions for WSS

Type of regulatory function	Definition
Tariff regulation	Setting and updating prices, determining tariffs by consumer group, establishing caps on revenues or the rate of return on investment.
Establishing quality standards for drinking water	Setting quality standards for drinking water and/or monitoring compliance.
Setting quality standards for wastewater treatment	Setting quality standards for wastewater treatment and wastewater discharges and/or monitoring compliance.
Defining public service obligations/social regulation	Setting public service obligations (including requirements on service access) and performance requirements for operators.
Defining technical/industry and service standards	Developing standards underpinning the technical modalities and level of service delivery.
Setting incentives for efficient use of water resources	Establishing incentives or specific schemes to promote efficient water resource use.
Setting incentives for efficient investment	Establishing incentives or specific schemes to promote efficient investment.
Promoting innovative technologies	Establishing incentives or specific schemes to promote innovative technologies.
Promoting demand management	Establishing incentives or specific schemes to promote reduced water demands.
Analysing water utilities' investment plans / business plans	In some cases, the regulator may be asked to approve utilities' business plan or investment plan.
Information and data gathering	Collecting data from operators, undertaking market research to identify trends and potential risks.
Monitoring service delivery performance	Monitoring the performance of water services against a set of targets or of performance indicators; this can involve benchmarking water utilities.
Licensing water operators	Granting or approving licences for the operation of water systems.
Supervising contracts with utilities/ private actors	The obligations granted by the public authorities to a specific utility may be detailed in a specific contract (it is usually the case when a private actor is brought in). The regulator may be tasked with the supervision of the contract.
Supervising utilities' financing activities	Monitoring water utilities' financial schemes (e.g. bond issuance, equity investments).
Carrying out management audits on utilities	Auditing and/or approving utilities' business plans.
Customer engagement	Consulting with customers on regulatory issues; communicating regulatory decisions to the public.
Consumer protection and dispute resolution	Handling consumer complaints about regulated entities.
Advice and advocacy	Providing advice for policy making and project implementation; identifying opportunities for reforms, encouraging improvements to the regulatory framework.

Source: OECD (forthcoming a), *The Governance of Water Regulators*, OECD Publishing, Paris.

References

Greater London Authority (2011), *London's strategy: securing London's water future. The mayor's water strategy*, Greater London Authority, London.

Kim, S.-J., Schumann A. and Ahrend R. (forthcoming), *What Governance for Metropolitan Areas?*, OECD Regional Development Working Papers, OECD Publishing, Paris, http://dx.doi.org/10.1787/9789264231092-en.

OECD (forthcoming a), *The Governance of Water Regulators*, OECD Publishing, Paris, http://dx.doi.org/10.1787/9789264231092-en.

OECD (forthcoming b), *Water Governance in Cities*, OECD Publishing, Paris.

OECD (forthcoming c), *Stakeholder Engagement for Inclusive Water Governance*, OECD Publishing, Paris, http://dx.doi.org/10.1787/9789264231122-en.

OECD (2015), *Governing the City*, OECD Publishing, Paris, http://dx.doi.org/10.1787/9789264226500-en.

OECD (2014), *The Governance of Regulators, OECD Best Practice Principles for Regulatory Policy,* OECD Publishing, Paris, http://dx.doi.org/10.1787/9789264209015-en.

OECD (2013), *Making Water Reform Happen in Mexico,* OECD Publishing, Paris, http://dx.doi.org/10.1787/9789264187894-en.

OECD (2011), *Water Governance in OECD countries. A multi-level approac*h, OECD Publishing, Paris, http://dx.doi.org/10.1787/9789264119284-en.

Glossary

Black water: black water is defined as wastewater that contains faecal matter and urine. It differs from grey water, which is wastewater that contains residues from washing processes.

Brownfield (project, development or redevelopment, water infrastructure): brownfield refers to development occurring on previously developed land, as cities are being continuously rebuilt. Brownfield contrast with greenfield development, which occurs on previously developed land. Similarly, brownfield water infrastructure refers to existing infrastructure that needs to be retrofitted to meet new requirements.

City: see *Urban area*.

Counter-party risk: counter-party risk is defined as the risk to each party of a contract that the counterparty will not live up to its contractual obligations. In most financial contracts, counterparty risk is also known as default risk (www.investopedia.com/terms/c/counterpartyrisk.asp).

Disruptive innovation: the OECD's Oslo Manual defines a disruptive (or radical) innovation as an innovation that has a significant impact on a market and on the economic activity of firms in that market. This concept focuses on the impact of innovations as opposed to their novelty. The impact can, for example, change the structure of the market, create new markets or render existing products obsolete (www.oecd.org/sti/oslomanual).

Evapotranspiration: water lost to the atmosphere from the ground surface. It is the combined process of evaporation of water stored on and below the ground surface, and the transpiration of groundwater by vegetation. (Source: OECD, 2014, *Climate Change, Water and Agriculture: Towards Resilient Systems*, OECD Publishing, Paris, http://dx.doi.org/10.1787/9789264209138-6-en)

Greenfield (area, development, project, or site): Greenfield developments are defined as the creation of planned communities on land not yet converted to development (typically on green surroundings of cities). They contrast with brownfield developments, which occur on previously developed land.

Grey water: grey water is defined as wastewater that contains residues from washing processes. It differs from black water, which is wastewater containing faecal matter and urine.

Indirect potable reuse: a process whereby purified recycled water is discharged into a water body before being used in the potable water system (www.oecd.org/env/resources/42349741.pdf).

Metropolitan area: a metropolitan area encompasses different municipalities that are socio-economically closely connected to the central city (or cities) or included in a functional urban area, representing the integration between (large) cities and rural surroundings. (Source: OECD, 2013, *Definition of Functional Urban Areas (FUA) for the OECD metropolitan database,* www.oecd.org/gov/regional-policy/Definition-of-Functional-Urban-Areas-for-the-OECD-metropolitan-database.pdf)

Membrane technology: membrane technology covers all engineering approaches for the transport of substances between two fractions (*permeate* and *retentate*) with the help of permeable membranes. In general, mechanical separation processes for separating liquid streams use membrane technology. Membrane separation processes operate without heating and therefore use less energy than conventional thermal separation processes such as distillation, sublimation or crystallization. Both fractions can be used (http://en.wikipedia.org/wiki/Membrane_technology).

Performance-based contract: performance-based contracts are developed to help define the development goals of the operator of water services. These contracts include time-bound performance targets against which the performance of the operator is measured. Unlike traditional government contracts that focus on inputs (procedures and processes to be used in delivering a service; amount and type of equipment; and/or time and labour to be used), performance-based contracts focus on results thus encouraging operators to be innovative and to find cost-effective ways of delivering services (www.oecd.org/env/outreach/47425194.pdf).

Urban area: acknowledging the limitations of definitions using thresholds based on population density, the OECD proposes a new definition based on economic function rather than administrative boundaries. In collaboration with the European Union, the OECD has developed a harmonised definition of urban areas as functional economic units. The definition has helped identify, for each OECD country, all urban systems with a population of at least 50 000 inhabitants. (Source: OECD, 2013, *Definition of Functional Urban Areas (FUA) for the OECD metropolitan database,* www.oecd.org/gov/regional-policy/Definition-of-Functional-Urban-Areas-for-the-OECD-metropolitan-database.pdf)

Urban metabolism: urban metabolism is defined as the study of the physical flows required to serve the urban economy (see Fernandez, J. E. (2014), *Urban Metabolism: City Typologies,* presentation at 2014 MIT Europe Conference, Brussels).

Water security: water security is defined as maintaining acceptable levels for four water risks: risk of shortage (including droughts); risk of inadequate quality; risk of excess (including floods); risk of undermining the resilience of freshwater systems. OECD work on water security claims that water management, at its core, is about reducing or avoiding water risks and about distribution of the water risks that remain (OECD, 2013, *Water Security for Better Lives*, OECD Publishing, Paris, http://dx.doi.org/10.1787/9789264202405-en).

ORGANISATION FOR ECONOMIC CO-OPERATION AND DEVELOPMENT

The OECD is a unique forum where governments work together to address the economic, social and environmental challenges of globalisation. The OECD is also at the forefront of efforts to understand and to help governments respond to new developments and concerns, such as corporate governance, the information economy and the challenges of an ageing population. The Organisation provides a setting where governments can compare policy experiences, seek answers to common problems, identify good practice and work to co-ordinate domestic and international policies.

The OECD member countries are: Australia, Austria, Belgium, Canada, Chile, the Czech Republic, Denmark, Estonia, Finland, France, Germany, Greece, Hungary, Iceland, Ireland, Israel, Italy, Japan, Korea, Luxembourg, Mexico, the Netherlands, New Zealand, Norway, Poland, Portugal, the Slovak Republic, Slovenia, Spain, Sweden, Switzerland, Turkey, the United Kingdom and the United States. The European Commission takes part in the work of the OECD.

OECD Publishing disseminates widely the results of the Organisation's statistics gathering and research on economic, social and environmental issues, as well as the conventions, guidelines and standards agreed by its members.

OECD PUBLISHING, 2, rue André-Pascal, 75775 PARIS CEDEX 16
(97 2015 05 1 P) ISBN 978-92-64-23010-1 – 2015-17

IWA Publishing's authorised EU representative for General Product Safety Regulations is Diane D'Arras, 15 rue Duret, 75116 Paris, France, e-mail: safety@iwap.co.uk.

Printed and bound by CPI Group (UK) Ltd, Croydon, CR0 4YY
12/05/2026
02108886-0014